JAMES MCCARTHY is the former Deputy Director (Scotland) of the Nature Conservancy Council and is a Board Member of its successor, Scottish Natural Heritage, representing that organisation as a director of the Edinburgh Green Belt Trust. One time Vice-Chairman of the Secretary of State's Working Group on Environmental Education, he is currently Chairman of the Environmental Youth Work Development Project Scotland, and a member of the Scottish Council for National Parks and of the Scottish Environmental Education Council. He is a member of the Forestry Commision's Mid Scotland Regional Advisory Committee.

Previously a forest officer in East Africa and widely travelled in Europe, Canada, USA and Australia, James McCarthy's 30 years in active land and conservation management at home and overseas enables him to take a wide perspective on his native country as environmental consultant, writer, lecturer and guide. He is the author of *Scotland:The Land and its Uses* (Chambers 1994), now revised, updated and expanded as *An Inhabited Solitude: Scotland, Land and People* (Luath Press 1998) and of articles on the wildlife of Scotland for Scottish Tourist Board's *Scotland: Where to Go and What to See*, among several other publications. He is committed to promoting Scotland as a unique environment which he believes is often under-rated, and to encourage sustainable tourism through holistic education, connecting natural heritage, history and culture for a deeper understanding of the country.

LAURIE CAMPBELL is one of Britain's leading wildlife photographers. He is the author of two books: *RSPB Guide to Bird & Nature Photography* and *The Wildlife Photographs of Laurie Campbell*, and has illustrated several more including *The Great Wood of Caledon*, *The Cairngorm Reindeer* and *Golden Eagles*. Forthcoming titles include a major project on badgers, and one on Patagonia.

Laurie has had 15 highly commended pictures, one specially commended picture, one category runner-up, and 3 outright category winning pictures in the BBC Wildlife Photographer of the Year Competition.

He studied photography at Napier University, working as a photographer at Edinburgh University department of medical illustration before becoming a freelance nature/landscape photographer, specialising in photographing Scottish landscapes and wildlife in the field using 35mm equipment and natural lighting whenever possible.

Laurie provided all the photographs for this book, and wrote Chapter 14 *Wildlife on Film*.

DUNCAN BRYDEN, who contributed the chapter on sustainable tourism, manages the Scottish Tourism and Environment Initiative and has been watching wildlife for over 30 years.

Wild Scotland

JAMES McCARTHY

Luath Press Limited

EDINBURGH

www.luath.co.uk

First published 1995 (Scottish Tourist Board)

Revised, extended and updated 1998

The paper used in this book
is produced from renewable forests
and is chlorine-free.

Printed and bound by Gwasg Dinefwr Press Ltd., Llandybie

Typeset in 9.5 and 9 point Sabon by
S. Fairgrieve, Edinburgh, 0131 658 1763

Contents

PART FOUR: PLACES TO ENJOY WILDLIFE

Foreword

LET ME DECLARE AN interest. I am writing this Foreword not only as chairman of Scottish Natural Heritage but also as a friend and colleague of the author. Jim McCarthy was the Deputy Director of the old Nature Conservancy Council in Scotland when I become involved in its affairs nearly ten years ago, and he is currently a member of SNH's East Areas Board.

But 'interest' does not necessarily mean 'bias'. My 'interest' is above all in Scotland's natural heritage, and this book does it proud.

It is a visitors' guide with a very distinct difference. We know from surveys that the vast majority of visitors to Scotland put 'the natural heritage' and 'scenery' right at the top of their reasons for coming to Scotland; this book concentrates attention on those very aspects of Scotland – not just for the bird-watching or botanist but for anyone with an interest in our incomparable countryside.

With his years of experience and inside knowledge as a naturalist and conservationist, Jim McCarthy makes a knowledgeable, unobtrusive and sure-footed guide and mentor across the length and breadth of Scotland and its islands. But the book does much more than provide a mere compendium of places which offer the best viewing facilities to see wildlife and plants and spectacular geological features – the special places like National Nature Reserves and the small reserves run by local authorities and voluntary bodies; it also takes us behind the scenery, as it were. It illuminates the diversity and character of Scotland's wildlife, the individual species of plants and animals which makes Scotland so special, and it tells of the various measures being taken, both by official bodies and by voluntary organisations, to safeguard and enhance the natural heritage, and to make it more accessible to the public.

Wild Scotland encourages visitors and residents alike to explore and enjoy the natural riches of our countryside in a responsible and informed way. It is a companion volume to *An Inhabited Solitude: Scotland – Land and People*, which is being published simultaneously. Together, they give everyone the chance to take the fullest possible advantage of what our wild Scotland has to offer to the discriminating.

Magnus Magnusson KBE,
Chairman, Scottish Natural Heritage

CHAPTER I

Wild Scotland

SCOTLAND IS DEFINITELY DIFFERENT – or so our many visitors tell us! When we ask what makes the difference, they tell us about the castles, the attractive small towns and villages, the friendly people – but invariably they enthuse about the unspoiled countryside and the beauty of our natural heritage, its mountains, lochs and coast. This is hardly surprising when we have one of the lowest population densities in Europe.

It can all be summed up in one word – diversity: from the rich cornfields of Berwickshire to the lonely islands of the north-west, and in between, majestic mountains and sweeps of heather moorland. It starts with the geology, which is among the most varied and ancient of anywhere in the world – landscapes which everywhere have been shaped and gouged out by ice thousands of years ago. The Highland Boundary Fault, which separates the two distinctive upland and lowland faces of Scotland can be seen at the thundering cataract of the Reekie Linn on the Angus/Perthshire border, or on the green jewels of islands marking the zone across the middle of Loch Lomond. Add to this a total rainfall which can drop by two thirds over a hundred miles and an ever-changing weather pattern, and you have a country where the land and the sky are constantly altering their colours and textures. The clarity of the night sky in rural Scotland may come as a revelation to eyes which have only experienced the light pollution of densely populated and illuminated areas.

Scotland's landscape is renowned, with National Scenic Areas covering a significant proportion of the country but there are also many other Areas of Great Landscape Value. These include not only the most spectacular mountain country, but also the gentle pastoral land of the lowlands. However, it is the typical highland scenery which dominates the image of Scotland's landscape. The arctic plateau of the Cairngorms, the wilderness of Wester Ross, and the woods and loch scenery of Loch Lomond and the Trossachs are not only outstandingly beautiful, but including many nature reserves, are also among the most important sanctuaries for characteristic Scottish wildlife.

To visitors from other countries, it comes as a surprise to find so much open country where there are quite dramatic views even at low altitudes. A largely undeveloped coastline, extending to 10,000km comes as another revelation: nowhere else in continental Europe outside Norway can you find anything quite like the long western sea lochs bringing the salt air of the Atlantic far inland. A multitude of unspoiled freshwater lochs are nourished by fast-flowing highland burns set in a green and well-watered countryside. Visitors do not have to go to special places to see the country's wildlife – it is all around them and access to the countryside is easy. Visitors who base themselves not too far from the Highland Boundary Fault, in one of the many attractive townships on the edge of the Highlands, have the best chance of seeing the greatest range of Scotland's wildlife, encompassing both highland and lowland aspects, within easy travelling distance, including a coast which is never far away – and a proportion of the 90,000 plants and animals (from the humblest microbe to red deer) estimated to occur within Scotland.

Scotland is a year-round destination for wildlife and scenery. From mid-autumn to early spring, there is the great spectacle of wintering wildfowl and waders flighting to and from the firths and estuaries, while those crisp late winter days are often the best to see deer at close quarters against a backdrop of snow-capped mountains. Indeed, there are particular opportunities for seeing wildlife in the short days of winter, with the limited daylight available to animals for activity, while their presence is often indicated by tracks in the snow. Those who are knowledgeable about Scotland choose the driest months of May and June to visit the west coast and islands with their great shows of coastal flowers and spring bluebells car-

Did you know that . . .

* Enjoying the countryside is one of Scotland's most popular leisure pursuits ?

* Scotland has 36 Country Parks which provide easily accessible country side in and around our major cities ?

* Around 70 million recreation trips are made to Scotland's countryside or coast every year ?

* About 60% of the population take at least one walk in the countryside each year ?

* Together, day trips for tourism and recreation generate about £200 million annual expenditure and 20,000 jobs for Scotland ?

* That the scenery of Scotland was rated as the top attraction by a considerable margin for all tourists ?

* That the top six species which visitors most want to see are seals, sea birds, birds of prey, dolphins and whales, red deer and otters ?

* Almost 30% of tourists to Shetland come to see its huge seabird colonies, generating more than £1 million in revenue ?

Sources: Review of Wildlife Tourism in Scotland - Tourism and Environment Task Force 1997
Facts and Figures 1996-97 – Scottish Natural Heritage 1997

peting the oakwoods of Argyll and Galloway. A little later is the time to see the mountain flowers at their best, and the plants of marshes and loch shores.

The explosion of breeding birds on the coastal cliffs starts in April and many of the coastal reserves are important staging posts for autumn migrants. Some visitors consider that late summer and autumn are the finest months in Scotland, the soft light bringing out the colours of red-berried rowan against the russet of bracken and purple heather. All are agreed that the scenery of Scotland, under a changing skyscape, presents dramatically different aspects to be enjoyed at any season. That scenery is the result of the interaction of geology and climate to produce the vegetation and wildlife which has been so powerfully modified by man in

---

Watching Wildlife

With few exceptions all wildlife including plants, are legally protected in Scotland: it is illegal for example to disturb birds on the nest, or to photograph nesting birds without a licence (advice on licensing requirements can be obtained from Scottish Natural Heritage). Unfortunately, over-enthusiastic naturalists and bird watchers can be the worst offenders! Uncontrolled dogs can cause considerable disturbance to many animals and birds.

* Avoid disturbance at nest sites - dogs can be a particular problem in keeping birds off their nests

* Do not linger at nest sites - remember that eggs can be destroyed by getting cold just as much as by breaking

* Watch for alarmed birds and move away

* Very cold weather is a testing time for birds - food and water are especially needed

* In areas where ground-nesting species are likely, watch where you tread!

* Do not leave dangerous litter

* Keep to paths if possible

* Protect bird habitats from fire

You are much more likely to get the best sightings if you:

* Sit quietly an wait for wildlife to come to you

* Use binoculars

* Avoid bright clothes

* Use observation hides wherever available

* Get out early

* Don't forget that train and ferry journeys offer some of the best opportunities for wildlife viewing

3

many places down the centuries. Although Scotland's landscape may look relatively natural, very little if any of our countryside has not been affected by human activity, so that the term 'natural' must usually be qualified: some 40 per cent of our flora and 18 per cent of our vertebrates are non-native.

John Muir - Scottish Founder of National Parks

The first national park system in the world was established in the USA at the end of the last century- a model which has been copied by many other countries since. What is less well known is that this came about largely as a result of the explorations, oratory and writings of a remarkable Scot, born in Dunbar, a small coastal town in East Lothian. John Muir spent his early childhood there before emigrating to America with his family in 1849. His first victory was in obtaining approval of the US Congress for the protection of the Yosemite Valley: now over 200 natural areas bear his name, while in Scotland he is celebrated in the John Muir Country Park, (see *Places to Enjoy Wildlife*) a wonderfully diverse stretch of coast adjacent to his home town. He has been described as the founder of the modern environment movement and the John Muir Trust has been established in his honour to acquire wilderness for conservation.

It is the same distinctive geology and climate, coupled with Scotland's geographical position which has created unique assemblages of species and habitats, now recognised as being of international importance. Thus Scotland has a number of species which are either at the southern or northern limits of their natural ranges, and in the case of many of our non-flowering plants such as ferns, mosses, liverworts and lichens, a number are subtropical and fragmented in their world distribution. Because of exposure, it is possible in walking up a Scottish mountain to cross a whole series of vegetation zones which might only be encountered elsewhere by travelling a thousand miles north! Some apparently common species, such as the heathers, occur over continuous and extensive areas not found outside Scotland, while our post-glacial history has ensured the survival of unique combinations of both arctic and alpine plant species. The purpose of this book is to provide a helpful guide for the interested general visitor to our countryside on the range of our natural heritage, and where to see it most easily, although it is hoped that it might also be useful to those who already have some knowledge of our wildlife.

CHAPTER 2

Geology and Landforms

THE FOSSIL RECORD TELLS US that more than 450 million years ago, Scotland lay south of the equator and was separated from England by an ancient sea almost a thousand miles wide. There were times when it was a desert, and others when it was covered by tropical rainforest. Sometimes the country sat under a huge thick sheet of ice. About 500 million years ago, Scotland was on the edge of a vast ancient northern continent which linked Scandinavia, Greenland and North America. Enormous sheets of eroded sands from this continent emerged to create the base rocks of grey and pink gneiss over much of northern Scotland. To hold in the hand a pebble of banded gneiss from the Atlantic shore of the western Hebrides and realise that it might have come from the coast of Nova Scotia more than 500 million years ago is a mind-boggling experience.

The last glaciation came to an end about 10,000 years ago. There have been at least five ice ages in the last three million years, and this ice, nearly a mile thick, has greatly changed almost every aspect of Scotland's landscape. The vast amounts of ice and rock dumped by the melting glaciers have altered the original courses of rivers, and deep lochs, such as Loch Lomond, were created by the gouging action of the glaciers. Mountain tops were sheared off, high-altitude ice lochs and corries were formed and valleys were widened and deepened. As a result of the submersion of the land under the weight of ice and its subsequent melting – causing the land to rise again – round much of Scotland's coast there are obvious second shore levels above the present beaches.

Extensive peat deposits now cover much of the north and west, the result of a cool, moist climate and widespread acid rocks which inhibit vegetation breakdown and soil formation – the peatlands of Caithness and Sutherland alone represent seven per cent of the total world resource of this formation. Although there are soils as fertile as any other in Britain, particularly those derived from the rich red sandstones along the east coast, these are localised. Elsewhere, the geology of the country often produces shallow sandy or organic soils with poor drainage, both of which are often low in plant food, interspersed with pockets of much richer soils developed from localised limestone deposits. Such nutrients as there are may be rapidly washed out by high rainfall, especially in the west, and be deposited as a hard impervious iron layer under the peat, preventing tree root penetration.

Given the size of the country, Scotland has an amazing variety of

5

rocks, as any glance at a geological map of the country will show. These range from the greatly altered Pre-Cambrian rocks in the north west, to the more recent volcanic lavas and the economically important marine sediments of the central Lowlands from the Carboniferous period. There are four distinct geological regions, separated by the major fault and thrust lines which cross the country, mainly from north-east to south-west. The most obvious is the Highland Boundary Fault, which separates highland from lowland Scotland, running from Stonehaven on the Kincardine coast to the island of Arran in the south-west, passing through the middle of Loch Lomond, where it shows up as a line of wooded islands. The highland area is divided by another thrust line 100 miles long from Loch Eriboll in Sutherland to the Sleat Peninsula on Skye, dividing the extreme northwest (including the Outer Hebrides) from the northern and central Highlands dominated by the Cairngorm massif. The southern side of the central Lowlands is marked by the Southern Uplands Fault, which is less obvious, although the appearance of these Border Hills to the south, on quite different geological strata, is distinctive. The Great Glen, in which Loch Ness sits, runs southwest from Inverness and near Fort William is penetrated by Loch Lhinne.

In each of these four regions, the dominant geology is very different. In the north west Highlands and Islands, we find the oldest rocks of all, the ancient grey Lewisian gneiss which has been thrust upwards from the lower earth's crust for a distance twice the height of Mount Everest, particularly

Did you know that . . .

* The rocks in Scotland are amongst the oldest in the world – about 3 billion years old ?
* Some of the rocks were pushed up from a depth equal to twice the height of Mount Everest ?
* Scotland was at one time part of a huge continent linking North America and Scandinavia ?
* Scotland was separated from England by a sea 1,000 miles wide ?
* Scotland was once south of the equator and basked in sub-tropical temperatures ?
* The fossilised trees which grew then can still be seen in Glasgow's Fossil Park ?
* Fife has the largest concentration of volcanic vents of any area in Europe?
* If the start of our geology is thought of as 24 hours ago, the last ice age (which finished 10,000 years ago) ended in the last second ?

well seen round the east coast of Harris. In the Torridon District, only slightly younger rocks, the red pebbly sandstones, were laid down by great rivers to depths exceeding five miles. Subsequent erosion has produced the fantastic isolated peaks characteristic of this district. On Skye, the geological landscape is again different, where the igneous gabbro rock has produced the climbing ridges and pinnacles of the Cuillins, while to the north, spectacular coastal cliffs are formed by the great lava flows of previous millennia. The island of Rum is also a magnet for geologists, with its unique volcanic and igneous complexes.

The strata of the Northern and Central Highlands are dominated by metamorphic rocks, altered by heat and pressure. These are mainly Dalradian and Moine schists and quartzites laid down between 600 and 1,200 million years ago and subject to enormous folding during the Caledonian mountain-building era. At Knockan Cliffs in the Inverpolly National Nature Reserve, there is an excellent demonstration of the rocks of the Moine Thrust well displayed in the geological trail. Great granite massifs such as the Cairngorms have been pushed under these strata, to be exposed later by erosion. It is in this area also that we find some of the most outstanding examples of glaciated country, such as the great U-shaped valley of the Lairig Ghru. This is in complete contrast to Caithness and Orkney in the extreme north of this region, where the sandstone flags are a distinctive feature of the countryside (in Caithness, especially where they are often used as walls between fields) and elsewhere erode into dramatic coastal cliff scenery and sea stacks off-shore.

Central Rum landscape

While the geology of the Central Lowlands is much less spectacular, the Carboniferous sediments, with their coals, mudstones, limestones, and even oil-bearing shales, have provided most of the important economic mineral deposits of the country. To the north of this belt, the deep rich red soils of Angus, Kincardine, and around the Forth are derived from the Old Red Sandstone series, creating some of the best agricultural land in Scotland. Extensive volcanic activity is indicated by such well-known landmarks as North Berwick Law and Arthur's Seat in Edinburgh, renowned as one of the sites where that great Scottish geologist James Hutton towards the end of the 18th century, demonstrated his Theory of the Earth. An excellent detailed geological trail within Holyrood Park published by Historic Scotland, Scottish Natural Heritage and the Edinburgh Geological Society guides the visitor around this classic site. Another such historic Hutton location is at Newton's Point near Lochranza on Arran, a west coast island which, straddling both highlands and lowland on either side of the Highland Boundary Fault provides a microcosm of the geology of Scotland – a "must" for anyone interested in the geology of the country.

Between the Central Belt and the English Border lie the Southern Uplands, mainly sedimentary shales and slates which produce the characteristic green rounded hills of the Border country, derived from the marine sediments between 400 and 500 million years ago, when Scotland was separated from England by the ancient Sea of Iapetus. The folding of these rocks can be seen most dramatically at St. Abb's Head in Berwickshire. Here again volcanic activity has created such prominent features as the Eildon Hills in the Middle Tweed valley, while on the border itself, the distinctive barrier of the Cheviot range is largely of lavas. To the west, slightly surprisingly, granite bosses punctuate the Galloway hills at Criffel near Dumfries and at the Cairnsmore of Fleet near Newton Stewart, creating an almost Highland landscape.

Scotland's natural history and geology come together when we consider its rich fossil flora and fauna. In the sandy flagstones of Caithness and Orkney we find examples of the aggressive armoured fish which preyed on the primitive creatures which were able to survive under the desert conditions of that time, some 400 million years ago. Some 100 million years later the muds laid down in the central belt supported giant millipedes and scorpions a metre long, while the fossil trees to be found in Fossil Grove, Glasgow, are evidence of the sub tropical climate of this time. In Caithness alone there are fossils representing no less than 60 species of sub-tropical plants, fish and fossilised corals. On Skye, fossils such as ammonites, gastropods, (snails), oysters, scallops, and worms are common in the Jurassic strata between Broadford and Elgol. On the other side of the country, the area around Dunbar with its limestones, has plentiful common fossils such as corals, crinoids ('sea lilies') and trilobites. Perhaps the most important recent finds have been the discovery of the world's earliest five-fingered creatures, parts of a crocodile about a metre long, and a tiny lizard, living over 300 million years ago in the Central Belt, when Scotland enjoyed an equatorial climate. These finds could be a vital pieces of the jig-saw linking the development of backboned animals with human evolution and the change from aquatic to life on land.

CHAPTER 3

Plants

FOR THOSE INTERESTED IN PLANT LIFE, Scotland is something of a revelation. On the higher mountains can be found rare species typical of the arctic growing at their southern limits, and alpine plants at considerably lower altitudes than in the Alps, all in a mixture unique in the world. These arctic-alpine plants include the lime-loving **mountain avens** growing from sea level in the north west, where the exposure is extreme, up to altitudes of 1,000m and more in the Cairngorms, and the **alpine gentian** with its intense blue colour on the slopes of the Lawers range above Loch Tay. Further east, on Ben-y-Vrackie near Pitlochry, can be found the delicate pale purple flowers of the **alpine milk vetch** and still further east, at the head of Glen Isla, the tall daisy like **alpine sow thistle** is found on rocky ledges.

The site descriptions under *Places to Enjoy Wildlife* make frequent reference to arctic-alpine plants, which as the name of the group suggests, are usually found at high latitudes much further north than Scotland or at altitudes well above those of our highest mountains, in both cases not far from perpetual snow where the mineral soil has been exposed. These conditions would have been widespread in Scotland 10,000 years ago as the glaciers retreated, allowing the first primitive soils to be developed, and their colonisation by those plants specially suited to withstand a cold environment. These circumstances are now only found at the higher altitudes in the Scottish mountains and the arctic-alpine flora is more diverse where lime-rich rocks are present and where the plants are protected from heavy grazing by their situation on cliff faces, ledges or the steepest ravines. Scotland's location and un-usual climatic regime has created montane environments where both arctic and alpine species occur in unique combinations, several of very rare occurrence.

Two examples of this historic group of plants are the alpine sow-thistle and **oblong woodsia**, both included in the Species Action Programme of Scottish Natural Heritage. The attractive alpine sow-thistle is confined to just a few inaccessible rock ledges at four sites in the eastern Grampian mountains. Outside Scotland, the species is found in the mountains of Europe, from Scandinavia to the Pyrenees and Carpathians, but here it is threatened by grazing and historic loss of mountain woodlands. By contrast, oblong woodsia is a relatively inconspicuous montane fern occurring at only three localities in Scotland, usually in rock crevices on high mountain crags or scree slopes. Its decline in Scotland has been catastrophic in the last 50 years, following earlier massive collecting in Victorian times,

9

The Seductive Invader

The visitor to Scotland in May or June is usually delighted to see the spectacular multi-coloured blooms of the several species of rhododendron which are advertised as a special feature of many gardens open to the public, especially in the west of Scotland, where both soil and climate clearly favour their growth. Introduced more than 200 years ago from Asia Minor, many different species have been added to these early introductions, which together with hybrids, offer a riot of colour in spring, from white, through red, to vibrant purple. The most common species is *Rhododendron ponticum*, which has the capacity, through its underground rhizomes, to spread at an alarming rate where conditions are favourable, especially in damp, shady, acid woodlands, including unfortunately some of our most valuable and scarce remnants of this particular habitat. Fiercely aggressive in its growth, ponticum can compete more than effectively with other native tree and shrub species, and casts a dense shade over the woodland floor, to the exclusion of most other ground flora. It will also invade wet bogland and once well established, it is a most difficult invader to eradicate, and thrives on cutting, to the despair of woodland managers.

and current conservation plans involve collecting spores and cultivation prior to translocation.

At the other extreme, at sea level on the western coasts and islands, we find a flora which is found nowhere else in Britain, notable not because of the rarity of the plants, but because of their sheer profusion. In these areas, lime-rich sand made up of countless millions of sea creatures underlies sea meadows of dune grassland known as *machair*. This is often lightly cultivated and hand fertilised with local seaweed for agriculture. In spring and summer it is a riot of colour, especially where the land has been left fallow. The eye is delighted by sheets of **bird's foot trefoil, kidney vetch, red clover, ox-eye daisy** among many others, while the wetter hollows are spectacular with yellow **flag iris** contrasting with masses of **ragged robin**, and the pink-tinged petals of the white **bog-bean** in the small lochans. Extensive *machair* can be found frequently along the west coasts of the Outer Hebrides, at Calgary Bay on Mull and on Tiree, Coll, Islay at Loch Gruinart, and on the mainland at Sanna Bay near the point of Ardnamurchan.

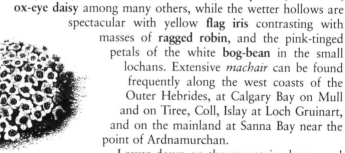

Purple saxifrage

Lower down on the mountain slopes, and especially in the eastern hills in late summer, the moorlands, such as those of Glenlivet and the Angus Glens, are ablaze with the purple **heather**, that emblematic plant of Scotland, often mixed with **blaeberry** (bilberry) and the black-berried **crowberry**. In the more open old Caledonian pinewoods, such as Abernethy, the salmon-pink bark and deep green of the **Scots Pine** set against the swathes of purple heather and grey lichens is a memorable sight. Tree and shrub species include the **downy birch** and the **juniper** with its dusky blue berries, and below you may find the **creeping ladies tresses** orchid and several species of the glossy-leaved **wintergreen**. Here and there, the emerald green of a wet bog will indicate deep **sphagnum** moss growth, often with the insect-eating **sundew** and **buttterwort** with its delicate purple flowers. In the west, especially around Loch Lomondside and in Glen Nant in Argyll, the moist and shady old oakwoods are renowned not only for their sheets of **wild hyacinths** (bluebells) in springtime, but also for the variety and luxuriance of their ferns and mosses.

Juniper

It is these non-flowering plants which characterise the dominant geographical element in the Scottish flora, which is described as Atlantic or oceanic, due to out north-western island situation. As previously indicated, Scotland can be divided into a number of zones based on the occurrence of certain indicator species. These are mainly plants,which may reflect our state of knowledge, since plant distribution is considerably better known than for example, insects. And of the plants, it is the relatively inconspicuous non-flowering groups – ferns, mosses, lichens and overwhelmingly liverworts – which enable us to differentiate such biogeographical regions. Although the arctic-alpine element of the Cairngorms, for example, is characterised by the presence of **alpine foxtail, alpine speedwell** and downy willow, the most strongly indicative plants of the West Highlands and Islands are several species of liverworts and mosses. By contrast, the eastern fringe of the Grampians and the Southern Uplands share such species as **mountain pansy** and **burnet saxifrage**, while the smallest of all the climatic zones, comprising only the islands of Barra and Tiree, are characterised by the occurrence of **sea fern grass** and **sea spleenwort**. Hardly surprisingly, with the prevalence of peat, the diagnostic plants of the moorlands of the Western Isles are species of Sphagnum moss.

The total number of plants of all groups (including algae, mosses, ferns and flowering plants) occurring in Scotland and the immediately surrounding sea is approximately 11,000. Of this, there are 1,117 native terrestrial plants (excluding the many arguable sub-species of Taraxacum and Hieracium). More than half of all the taxa in the flora of Scotland are derived from introductions. Two of Scotland's rarest plants are now threatened with extinction as a result of collection throughout their world range – **Killarney fern**, now known only from isolated sites in Arran and Argyll, and **Norwegian mugwort** found in the mountains of the north-west.

Scotland has nine endemic plant species – that is throughout the world, found only in Scotland – including **Scottish primrose**.

Probably the most attractive of our endemic plants *Primula scotica* or the Scottish primrose, is a most easily overlooked flower. A tiny plant, normally a mere 2.5 cm high growing from a rosette of short spear-shaped leaves only a few centimetres across, it is easily distinguished by its delicate purple flowers with a central yellow eye. It is found only on windswept headlands and dunes of the northernmost coast of mainland Scotland and some Orkney islands, usually in sandy grasslands or on cliff tops, flowering twice, in May and July. Although it is quite a striking plant, with its attractive bright flowers, it is often quite difficult to see, hidden among other plants or stones. Some sheep or rabbit grazing may be essential to restrict other competing vegetation and to allow this minute plant to survive. However, despite their small size, individual plants have been known to live for at least 20 years. Land use changes, especially agricultural improvement over the last 40 years have resulted in marked changes to its habitat and serious losses of a number of populations in recent years. Because of this, it is the subject of special conservation measures, including introductions to suitable habitat, as part of the Species Action Programme of Scottish Natural Heritage.

Scottish primrose

Within Britain, Scotland is particularly important botanically for its range and abundance of mountain, northern woodland and aquatic plants – of the plant species which within the UK are restricted to Scotland, some 45 are mountain plants, many from the Southern Highlands, while a further 9 species are coastal. Some 98 percent of plants which occur in Britain are to be found in Scotland, including 76 species which are exclusive to Scotland. The native pinewoods are also of special importance for their specialised flora. Scotland is unusual, in a European context, in the sheer extent of, for example, its spring bluebell displays and late summer heather – elsewhere in Europe, these are much more restricted. Heather *(see Bonnie and Purple)* is a good example of a species which is most abundant on the maritime fringe of Europe.

On the other hand, many of the plants previously associated with lowland farmland have declined dramatically this century, including species such as the attractive **corn cockle** (a relative of garden pinks), **cornflower** and **fumitory**, all regarded as agricultural weeds. With the intensification of agriculture, many such plants have largely disappeared or are restricted to a few relicts of uncultivated rich meadows, or even on road verges protected from the early roadside cutting which threatened their survival. Many

Peat and Pollen

The lack of oxygen in peat, which is plentiful in Scotland, makes it a good preservative – everything from prehistoric canoes to pollen grains. The latter fortunately are individually identifiable under the microscope, which makes them of enormous value to plant palaeontologists who are interested in investigating the plant history of the country. Bogs throughout Scotland have yielded important evidence of the distribution of plants over thousands of years, especially tree and shrub species, as a result of analysis of the deep peat cores which have been extracted from these living archives of our ecological history. Through carbon analysis, it is possible to date the peat layers relatively accurately, along with the changes in vegetation represented by the 'pollen rain' which fell into the wetlands of that time. These changes reflect not only alterations in climate – by the presence of different proportions of pollen grains over time – but also early man's influence, with the appearance of the first introduced plants, including cultivated cereals. Other reasons apart, this is one powerful justification for protecting these important reservoirs of our past natural history in the peatlands of today.

of the plants of the lowlands of Scotland are not especially different from those elsewhere in Britain, including the very many species which grow in urban situations or on waste land. A recent publication on the flowers of Glasgow describes over 60 native species (including 3 orchids) and there of course many more, greatly encouraged by that city's many fine parks where an enlightened attitude prevails towards leaving wild areas to look after themselves.

Scotland has been noted for its botanists for several centuries. In the 17th century such pioneers as Sibbald and Balfour, like many others, were medical practitioners and therefore interested in plants for their medicinal properties. Both were instrumental in founding the physic garden (initially near Holyrood) in 1670 which eventually became the renowned Royal Botanic Gardens in Edinburgh. They were followed by the first professors of Botany at Edinburgh University, including James Sutherland in the late 1690's and John Hope in 1761. These energetic men and their students travelled widely in Scotland to find new species of native plants, especially from the mountain districts, despite the difficulties of travel at that time; it was Hope in particular who effectively established botany as a separate branch of science from medicine. Subsequently Thomas Pennant's tours in Scotland became well-known through his prolific writing, and included many references to vegetation, such as his notes on the great pinewoods of the north.

One of Hope's students, Alexander Menzies, was appointed naturalist on the Discovery during Captain George Vancouver's voyage round the world in the 1790, but he was only one of several notable Scots botanists who travelled to some of the remotest parts of the Earth to bring back collections of plant seeds for both commercial and ornamental purposes, right up to modern times. David Douglas is celebrated in the fir of that name, which along with many other conifer species such as the now ubiquitous **sitka spruce**, he discovered in the North American Pacific Coast range in the 1820s. One of the very first naturalists  to consider plants in relation to the conditions in which they grew, which subsequently developed into the modern study of ecology, was William MacGillivray, who became Pro-fessor of Civil and Natural History at Aberdeen and was renowned for his hardiness and fortitude, even walking all the way to London in 1819 simply to see the bird collection at the British Museum! George Forrest, who died as recently as 1972, distinguished himself as a plant collector in China, where he made seven expeditions between 1904 and 1932, and was responsible for the introduction of many new species of rhododendron (see *The Seductive Invader*) to Europe. The tradition of botanical investigations in South-East Asia is strongly maintained by the present day Royal Botanic Gardens in Edinburgh.

Marsh marigold

Some of the results of the labours of these botanists can be seen in the many fine gardens and plant collections of Scotland, including the government-administered national gardens in Edinburgh, at the Younger Botanic Garden at Benmore in Argyll (see Argyll Forest Park under *Places to Enjoy Wildlife*) and at Logan Gardens (both extensions of the Royal Botanic Garden in Edinburgh) in Wigtownshire, with its many Southern Hemisphere species flourishing in one of the mildest climates in Scotland. Inverewe in the north-west is internationally famous as a garden which was created in the 19th century from the most unpromising exposed conditions in this west Highland situation. Under Scotland's Garden Scheme over 300 private gardens are open for charity throughout the country, including a number which are managed by the National Trust for Scotland, who are responsible for the internationally-recognised gardening school at Threave. Although of course many of the plants in these collections are exotics, they also contain a range of native trees, shrubs, and flowering plants and several are increasingly valuable refugia for threatened plants and a potential source for cultivation and reintroduction of these. The Dawyck Botanic Garden near Peebles for example claims to have the first Cryptogamic Sanctuary and Reserve for non-flowering plants in the world.

'Wha daur Meddle wi' Me'

Alongside the **rose** of England and the **shamrock** of Ireland the **thistle** is generally acknowledged as the plant emblem of Scotland – though perhaps others might suggest less kindly, acknowledging the Scots' reputation for prudence in money matters, that the great drifts of the purple **thrift** along the Scottish coasts would be an appropriate symbolic Scottish plant! However, while the **bluebells** of Scotland or 'witches thimbles' have been sung about, it is in fact the thistle, used by the early Scottish kings in their heraldry, which has come to formally represent Scotland in various armorial bearings, with the Latin inscription *'Nemo me impune lacessit'* replacing the vernacular challenge above, and adopted appropriately as the motto of the Scottish regiments. It has also given its name to one of the oldest Royal orders of chivalry in the country. In some ways it may also be

---

Gaelic Plant Names

While Gaelic plant names often correspond to their equivalents in English, many are quite different – frequently very descriptive, poetic, or even mysterious. What is one to make of 'The Waistbelt of Cuchullin', the legendary Celtic hero, who has given his name to **meadow sweet** or *'Cneas chù Chulainn'*, and why are all the **stitchworts** 'dejected'? (*Tùrsach/Tursarain*). It is easy to see why the insectivorous **great sundew** might be 'Lus a ghadmainn' or 'plant of the nit insect', but why is the **heath milkwort** 'fairy women's soap'? (*'Siabann nam Ban-sidhe'*). Fairies crop up again in the **bulrush**, not surprisingly as 'fairy women's distaff '(*'Cuigeal nam Ban-sidhe'*) Nothing could be more descriptive of its texture than 'mountain silk' (*'Sioda Monaidh'*) for **harestail cottongrass** or 'small polished one' (*'Liobhag Brochaich'*) for the **bog pondweed**, nor the charm of 'Reul na Coille' or star of the wood for **chickweed wintergreen** or 'pleasant little white one' (*'Fionnan Geal'*) for **grass of Parnassus**. On the other hand, 'blubber lip of the stream' (*'Meilleag an Uillt'*) makes the showy yellow **monkey flower** sound positively repellent! But the 'little white rose of Scotland', (*'Ròs Beag Bàn na h-Alba'*) celebrated in McDiarmid's' poem, is a definite improvement on the more prosaic Burnet Rose.

Source: Scottish Wild Flowers Michael Scott 1995

---

claimed to typify the Scottish character – prickly and proud, as in the French aphorism 'Fier comme les Ecossais!' There is the tale, no doubt apocryphal, of how the plant came to be chosen as the emblem of the country, when Norse invaders, in an attempted surprise attack, gave their presence away treading painfully on the thistles strewn in their path by the Scots defenders!

There are several species of thistle, one of which, the **field thistle**, is regarded as a serious pest of agricultural land. The **spear thistle** has much larger heads and is probably the one which is used now as the emblem of Scotland. Thistles grow in a wide variety of situations from marshy ground to dry farm fields. Some, with almost yellow flower heads such as the **carline thistle** (and therefore hardly conforming to the common notion of a thistle at all) are relatively rare, while the **melancholy thistle** – so named because of its solitary flower head – with its lack of sharp spines, could not be mistaken for the stereotype. This is perhaps best represented by that exuberant ornamental exotic, the **cotton thistle**, often two metres high and with distinctive silver leaves, to be found in castle grounds, and for example, in Princes Street Gardens in Edinburgh.

Bonnie and Purple

Apart from the emblematic thistle, no plant is more associated with Scotland than the heather, which imparts that wonderful colour to the late summer moorland. There are in fact three species: the early blooming **bell heather** *(Erica cinerea)* with its distinctly larger bell-shaped flowers, as much red as purple and which appear in July; **ling** *(Calluna vulgaris)* which is dominant on many Scottish moors later in the year; and finally, **crossleaved heath** *(Erica tetralix)* which is similar in appearance to bell heather, but is distinguished by being hairy, with larger flowers, and usually confined to wetter ground. Mixtures of all three frequently occur. Generally, heather prefers well-drained acid soils such as those of the Angus Glens and Grampians, and does not thrive so well in the west. Bell heather appears to do best on the sunny south-facing moors of the east.

Heather is quite crucial to red grouse, which feed on the young buds and also use the shrub for cover and nesting. The birds do best where there is a range of different aged heather close at hand, and to achieve this, the moors are burnt in spring on a cyclic pattern, ideally to create relatively small strips or patches at different growth stages. The burning is a specialised operation which requires to be carefully timed in relation to the weather conditions, if the burn is to be at just the right temperature to encourage regeneration of the young shoots, but not so fierce as to set back the plants, or worse, to lead to soil erosion, especially on steep slopes. Wind conditions are critical and with increasing extension of forestry, a heather fire which gets out of control can have disastrous consequences. While heather can provide essential nutrition for hill sheep especially when grass is unavailable in hard weather, intensive grazing will in time convert heather to grassland. A similar effect to burning can be produced by wind which maintains low growth heather in the most exposed coastal situations.

Heather has declined quite dramatically in recent years, because of the encouragement previously given to agricultural reclamation, with fertilising and seeding of hill land, but even more so as a result of blanket afforestation of the lower moorlands over the last 50 years. In that time, in regions such as Aberdeenshire, this has reduced the area of heather by over half, while in the lowlands, substantial areas have been lost to commercial peat extraction. Heather is also important to those beekeepers who specialise in heather honey, which connoisseurs claim to be superior to all other kinds, with that fragrance which can only come from the open summer moorland. In the past heather was valued for thatching and bedding, and mixed with peat, it provided strong walls for the crofts.

Heather moorland is not only important for grouse, but also for many other species which prey on this bird, such as golden eagle, peregrine and hen harrier, for animals such as blue hare, together with a whole range of insects dependent on the plant, including unfortunately, that serious pest, the heather beetle.

CHAPTER 4

Animals

COMPARED TO THE NUMBER OF terrestrial invertebrates in Scotland (probably over 20,000 and including over 14,000 insects), the number of mammals, amphibians and reptiles, at just over 240, is relatively small, yet it is these animals which capture our attention. Although the visitor is likely to see only a tiny fraction of our wild animals, those which are especially associated with Scotland will give a particular delight. Scotland lost many of its animals in prehistoric times, but it has still retained a rich fauna, including a number of game animals such as the **red deer** and grouse which are much prized by hunters, while the wild Scottish **salmon** is still regarded as the king of fish.

The shy **otter** requires much patient observation before it will reveal itself, while a herd of shaggy **wild goat** (more correctly termed feral goats since they are derived from earlier domesticated goats) will graze unconcerned a few yards away. There are about 2,500 such multi-coloured goats distributed over some 140 sites – the fights between competing billies in autumn is exciting. While speaking of goats, it is appropriate to mention **Soay sheep**, the most primitive form of domestic sheep surviving in Europe – no other sheep like it exists today, although it closely resembles the wild sheep found in the remoter mountain areas of Southern Europe and central Asia. It is still a pure strain, and is frequently seen in parks, away from its original natural habitat on St. Kilda (their ancestral home is the island of

Did you know that . . .

* In past times Scotland had native reindeer, lynx, beaver, wolf, wild boar, and bear?

* The last wolf was killed in Scotland in Morayshire in 1743 ?

* The limestone of the Inchnadamph Caves in Sutherland preserved many of the remains of these animals ?

* The Scottish bear was taken to the Roman circuses because of its ferocious reputation ?

* Many of these animals now extinct in Scotland are represented on the Pictish symbol stones in Eastern Scotland ?

* Most of them can be seen in the flesh at the Highland Wildlife Park at Kincraig near Aviemore ?

Soay in this archipelago) where there have been major fluctuations in its populations in recent years.

Almost the whole of the British population of **blue mountain hares** is to be found in Scotland. It is the only truly native member of the hare and rabbit family in this country and is restricted in its range due to competition from the introduced brown hare on lower ground. On the higher moorland, particularly in spring when the snow has partially melted, the mountain hare in its winter white coat may be quite conspicuous and is especially common in the eastern Cairngorms and the hills of Angus.

This is also the time when the red deer (see *Monarch of the Glen*) are seen in greatest numbers, before they have moved up the mountainsides with the onset of warmer weather – and flies! In the area west of Braemar and Deeside, as well as in Glenfeshie in Speyside, large herds of several hundred animals, both stags and hinds, can be seen on the lower hill slopes and valley bottoms. Set against the landscape of misty mountains, the red deer has become, from the popular pictures of Landseer onwards, a majestic symbol of wild Scotland. What is less well known is that it has only relatively recently become the light fleet-footed hill animal that we know today, when its true home is the forest – in European countries, the same species in its natural forest habitat is a much heavier, slower beast. In autumn the roaring

Wild goat

The Vendace – An Ice Age Fish

When the last glaciers melted about 10,000 years ago, they left behind lochs at both high and low altitudes which were characteristically arctic in type, and in which a number of typical arctic-alpine species survived. One of these is the vendace, a whitefish similar in appearance to herring, which has only been recorded in 4 sites in Britain, in the Lake District and in Dumfries-shire. Now the original populations (which were heavily fished in the 19th century) are thought to be extinct in Scotland, since there have been no catches in recent years from their last known locality, Mill Loch, Lochmaben, near Dumfries. Although vendace are widespread in Northern Europe, particularly Scandinavia, they have in Britain been threatened by changes in the chemical composition of their previous localities, especially by enrichment from sewage or agricultural fertilisers and the possible predation by other species. Plans are in hand to reintroduce the species to loch systems most likely to be unaffected by these changes (see Loch Skene/Grey Mare's Tail under *Places to Enjoy Wildlife*). The vendace is a good example of an historic native species which is highly vulnerable to changes in land use and to changing populations of competing fish species, sometimes as a result of deliberate introduction for sport fisheries.

of the stags reverberating round the glen during the rut is a heart-stopping sound. Red deer can also be seen frequently from the road across Rannoch Moor or on the A9 at Drummochter. The genetic integrity of the native red deer stock is now under threat from the successful interbreeding with the Sika Deer introduced to several deer parks over 120 years ago and now estimated to be as many as 10,000 animals, occupying 35 per cent of the red deer range.

The much smaller **roe deer** which inhabits mainly deciduous woodlands and especially young conifer plantations throughout the country gives itself away by the short almost dog-like bark of the buck demarcating its territory – frequently seen in the morning and evening with its distinctive white rump patch, feeding in fields surrounding woods. Since they were once part of our native fauna, mention should be made of the herd of **reindeer** which are still maintained at Glenmore in the Cairngorms. Although there is reference to hunting of this animal in Scotland in the 11th century, it seems more likely that reindeer became extinct in this country around about the same time as the appearance of man 9,000 years ago. The initial 18th century attempts at re-introduction were unsuccessful, and it was not until the 1950's that a small herd from Swedish Lapland, now numbering about 100, was brought into the Highlands, where they are a considerable tourist attraction.

Roe deer

Muntjac deer were first introduced into Britain in 1894, and have been seen in and around Inverness in recent years. Although smaller than the other native deer species, it is destructive, and there are fears that if its increase is allowed to go unchecked, it could pose a threat to woodland communities.

It will be a lucky visitor who spots a **wildcat**, now largely confined to the Scottish Highlands, although it has recovered considerably in numbers from previous persecution – in the mid 19th century there are reliable records of almost 200 wild cats being killed on one estate alone over a period of 3 years. A very rough estimate is of about 3,500 animals at present. It is a nocturnal hunter, and is most likely to be seen at first light by an observer with the patience to sit and wait at a suitable site, such as old woodland with rock outcrops and scree on the middle slopes of the hills. The cat will be recognised by its striped body and distinctive short ringed bushy tail, while the ears tend to be flattened outwards. However, there is now considerable doubt as to whether the so-called wild cat is a distinct species, as it is known to mate with domestic cats – with hybridisation between the two, wild and feral cats are now regarded as the same species.

The same habitat is also the haunt of the **pine marten**, a mainly noc-

A Natural Dam Builder

With a rich waterproofed fur and webbed hind feet, the European beaver is well adapted to its cold aquatic environment, feeding largely on grasses, herbaceous plants, leaves, and the twigs and bark of trees associated with rivers and lakes. Beavers appear to happily co-exist with human settlements and are easily observed: in several European countries (Finland, Sweden, Germany and the Netherlands) they have become considerable tourist attractions, partly because of their constant activity. Up to 400 years ago, this rodent was relatively common on the country's wetlands, but the trade in beaver pelts proved to be fatal to its survival here, as elsewhere in western Europe. Best known as fellers of trees (up to 40 cm in diameter) beavers can considerably modify their immediate environment by their dam-building to provide secure holts – in the process often creating habitats for other aquatic plants and animals which are favoured by fluctuating water levels. Compared to the reactions generated by suggestions to introduce the wolf, the more positive responses to the current proposals to reintroduce the beaver augurs well for the restoration of this attractive animal to Scotland's limited list of native mammals.

turnal animal which prefers woodland, especially with cliffs and scree. It is our rarest terrestrial mammal, readily identified by its creamy throat patch against otherwise dark brown fur, its bushy tail and its rapid bounding run. Like the wildcat, it is spreading with the increase in forestry and has taken to raiding dustbins from Beinn Eighe to the Great Glen south west of Inverness (where there have been spectacular increases in population) – making it more likely to be seen than the wildcat! Although it is now protected, it was at one time fiercely persecuted – it will of course take the eggs and young birds, whether domestic or wild.

The **red squirrel** has its main stronghold in Britain in our native pinewoods, where three-quarters of the total population occurs and where, unlike the pine marten, it is not generally increasing. The species became extinct in the lowlands of Scotland and was re-introduced there in the latter half of the 18th century. The evidence of red squirrels is easy to spot in the number of stripped cones littering the forest floor. From then on, it is simply a matter of keeping an eye open for the swaying pine branch or rapid scurrying movement round the tree trunk, especially very early in the morning in the old Caledonian pinewoods of Speyside and Deeside. There is now some concern that the increasing planting of deciduous woodland may encourage the expansion of the non-native grey squirrel at the expense of the red squirrel, which also appears to be seriously affected by a lethal virus to which the grey squirrel is immune.

Although numerous in some localities, the omnivorous **badgers** are thinly distributed in the Highlands compared to the south of Scotland and absent from the islands, with total numbers estimated at around 25,000. Often active in the winter, they are most likely to be seen at this time, especially with foliage absent. Badgers tend to favour woodlands in rolling country, avoiding wet areas, but somewhat surprisingly, can be found up to 450 metres altitude, where they will often have their dens between rocks. This is one animal most likely to be seen by being guided to a known sett in good time before they emerge. *(Check with the Scottish Wildlife Trust for their current badger watching services).*

Among the three native reptiles (the others being the common lizard and the slow worm), the **adder** is the only poisonous one, found mainly in south-west Scotland and the central Highlands, while it is absent from much of the Central Belt, the Outer Hebrides and the Northern Isles. Elsewhere, there has been a contraction of its range with a decided drop in numbers, especially on farmland. The dark zigzag pattern on the back is diagnostic, but adders, like other snakes, give humans a wide berth whenever possible – the likelihood of being bitten is usually when they are trodden upon inadvertently. Both the adder and the **common lizard** favour the drier moorland, especially in places where they can sun themselves on sandy soils.

There are now thought to be as many as 23,000 **red foxes** in Scotland, and these are frequently seen, especially during the winter months, in broad daylight. They occur virtually everywhere (other than the islands), from their rocky dens in the north-west highlands to underneath the author's garden shed near the centre of Edinburgh!

It is estimated that there are almost twice as many **stoats** as **weasels** in Scotland, so you are more likely to see the latter, easily distinguished

Natterjack Toad

One of the most threatened amphibians in Europe (it is now listed in the European Commission's Habitat Directive) this quaintly named toad with the distinctive yellow stripe running along its back, is sparsely distributed in Britain with just over 40 known colonies. Once described in its main North England location as the 'Southport Nightingale' from their evening chorus, the natterjack's loud croaking call can be heard for up to a mile on warm, still spring evenings. The most northerly breeding sites in Europe are to be found in four locations on the Scottish side of the Inner Solway. Here they now make up about 10-20 per cent of the total British population, although as recently as the 1970s they had dwindled to a few pairs, as a result of pollution by cattle faeces and loss of breeding ponds. Their significant recovery since then has been largely fortuitous: when the Wildfowl and Wetlands Trust created sandy screening barriers from shallow scrapes which filled with water on the margins of the Caerlaverock salt marshes for observing wildfowl, (see *Places to Enjoy Wildlife*) they had incidentally provided the ideal breeding conditions for the natterjack, with the equivalent of sand dunes and wet hollows nearby. Combined with control of domestic stock to reduce disturbance to the birds, this was just what the natterjack needed - now the four colonies on the Solway coast numbers several hundred healthy adults.

by the very sharp difference between the russet brown fur on the back contrasting with the creamy-coloured underside, and in winter, may turn completely white, although the distinctive black tip to the tail (absent in the weasel) is always retained. The stoat is not found on the Outer Isles nor on any of the small isles, while the weasel, our smallest carnivore at only 17 cm for the female and 20 cm for the male, is absent from all of the islands except Skye and Raasay. Weasels are much less likely to be found in the mountains than stoats, and almost never change colour in winter. While the stoat is found in all sorts of habitats (provided there is good cover) the weasel is more often found on lowland farmland where there are hedges and dykes.

Red fox

Both animals have the endearing habit, being curious animals, of sitting upright to get a better view and both are most likely to be seen on roadside verges or crossing the road. Although very much different in its almost black colour, the mink might at first be confused with stoats or weasels, especially since the escaped varieties are often initially much lighter in colour.

The **North American mink** were originally brought in to Britain for fur-farming, many of these establishments having long since closed down. However escapes have ensured that this fierce little predator has now unfortunately become well established in Scotland, where it has decimated bird colonies. This is particularly serious for the vulnerable ground nesting species of the islands, where like the introduced **hedgehog** (see *Vulnerable Islands*) it has run riot. However, where control programmes have been run, breeding of such fickle species as terns can be highly successful, even where as in Loch Etive, they had been absent for many years.

Because of the laws which now give protection to **bats**, they are more frequently in the news than previously. In Scotland, 8 native species breed, the commonest being the tiny **pipistrelle**, which can be seen in any month in the year, although mostly between March and October. Large numbers roost together, and they can be a nuisance to householders because of their accumulated droppings in attics, but like the other species in this country, are entirely harmless, and may even be considered beneficial in preying upon insects, notably midges. Considerably larger is **Daubenton's bat**, which like the **long-eared bat** is not found north of Inverness, and is often seen hunting over water, feeding almost exclusively on aquatic insects including the larger ones such as dragonflies and moths, and may therefore sometimes be caught on fishermen's flies. They will lightly dip on to the water surface to drink on the wing. The long-eared bat is distinguished by ears which are almost as long as its head and body together, and feeds primarily on butterflies and moths, as well as other insects. Finally, the **natterer's bat** is found from the Central Belt southwards, invariably associated with woodland and wooded parkland where some water is available. There are indications that because of climatic change, bats in Scotland may be extending their range here.

Monarch of the Glen

Although the **Great Irish Elk** (more properly **Giant Fallow Deer**) disappeared in prehistoric times, **Scandinavian elk** remains have been found in the ancient circular stone fortresses or *brochs* in the northern districts of Scotland. The Romans described the elk and the specific Gaelic name of *boirche* (as distinct from *feidh* meaning smaller deer species such as red deer) and many highland legends testify to the existence of this large, distinctively dark-coloured deer which probably survived into the 13th century. In 1781, on the Boroughmuir, only a mile from the centre of Edinburgh, a prehistoric red deer head was unearthed, with one many-pointed antler measuring almost 1.5m long. Other complete preserved carcasses indicate that the ancient beast would have been at least a third larger than its modern counterpart. If the deer gain access to the new forest plantations, as

Bring Back the Wolf?

The justification for referring specifically to an animal no longer in Scotland is that it is the only last known predator capable of hunting deer, and because there are frequently calls for its re-introduction - despite the reputation it has acquired for itself, however wrongly. There are many historical references to the wolf in Scotland from earliest times, including place names associated with it. In mediaeval times, because of the perceived threat from wolves, which was undoubtedly a menace to unprotected domestic stock, they could be hunted by anyone outside the royal forests - wolf hunting was a favourite pastime of royalty and nobles, and there is a record of five animals being killed in a hunt organised for Mary Queen of Scots in 1563. Wolf hunters received substantial bounties on the heads of wolves killed, and shelters known as 'spittals' were specially constructed to give sanctuary to travellers against the threat of wolves. The claimed grave-robbing tendencies of wolves aroused the greatest horror in earlier times. Despite a more enlightened attitude now, there are considerable difficulties associated with re-introduction, setting aside the understandable fear of this effective predator and its potential impact on livestock and game. Notwithstanding the excess of deer in some localities, the natural forest environment and the pattern of human occupation has changed considerably since the time when wolves roamed freely over Scotland, so that it is perhaps not surprising that enthusiasts for bringing back the wolf have not so far been able to in mount a convincing public campaign for re-introduction. The successful legal challenge by ranchers to the re-introduction of wolves to Yellowstone National Park reflects some of the problems of such operations.

they have for example in Galloway, they may again become more like the heavy woodland deer of the Continental forests.

On the island of Rum – now a National Nature Reserve – some of the most significant investigations on the ecology and behaviour of red deer have been carried out in recent years. Research there has shown that where a population is left unculled, the survival rate of individuals – especially the older stags – falls off sharply as the population rises. People have taken the place of previous predators, such as wolf and lynx, and among the archaeological finds on the island have been thousands of microliths, stone flakes made by the Mesolithic hunters which would have been attached to arrows or other weapons to hunt deer. In turn, deer antlers were amongst the commonest materials for making barbed arrows and other tools of the hunt. The kitchen middens from the Bronze Age onwards reveal plentiful remains of red deer. The disappearance of the forest, and persecution by man of the **wolf** (see *Bring back the Wolf?*), removed the red deer's only natural predator in this country in recent centuries.

Previously hunted with dogs, and trapped either in valleys which could be closed off or driven into specially constructed *tinchels* or enclosures, large numbers were trapped and killed. There is a record of James V with 12,000 men killing 18 score of deer in Teviotdale in South Scotland in one hunt, while further north, some 2,000 men could drive the equivalent number of deer in the great Forest of Athol. (Nowadays the term 'deer forest' still marked on Ordnance Survey maps is somewhat misleading, since the country is usually bare of trees other than recent plantations). There are vivid descriptions of these great hunts by kilted clansmen in the writings of travellers up to the 18th century. It was not until the early 19th century that the Duke of Bedford instituted the present method of stalking to hunt individual animals. Subsequently the publication of *Days of Deer Stalking* by William Scrope and illustrated by Charles and Edwin Landseer had enormous influence on the promotion of the sport. This form of hunting became highly fashionable, especially after its adoption by royalty, and led to the creation of the large sporting estates which are so much a feature of highland Scotland.

With the spread of sheep farming on a large scale in the Highlands from the late 1700s onwards, deer numbers fell correspondingly as the land was occupied by the Blackface and Cheviot flocks. Deer numbers again rose after 1825 with the decline in sheep as Empire sheep products were increasingly imported, and in 1860, for the first time, the rental for sporting estates exceeded that for sheep runs. In 1883 there were just under 1 million hectares of deer forest, which rose to a peak of over 2 million hectares by the outbreak of World War II. Today the annual cull is approximately 60,000 beasts, including stags, hinds and calves. Unfortunately, a combination of circumstances has led to population increases in red deer (from 15000 in the 1950s to well over 300,000 today), which prevent woodland regeneration and cause damage to other upland habitats. With deer now inhabiting more commercial forests and spreading into southern parts of the Kintyre and Cowal peninsulas and north Galloway, new culling targets have been set by the Red Deer Commission at 70,000 per year, up from the current level of 55,000.

The legends of the deer are of great antiquity and have passed down through the symbols and heraldry of Scotland. The coat of arms of the Burgh of the Canongate in Edinburgh depicting the stag's head, with a cross between its antlers representing the cross which David I saw and grasped to save his life when threatened by an attacking stag is one example. A number of highland regiments and Scottish clans (such as the Gordons, Frasers, Keiths and others) have a stag in their family arms. In all of these symbols it is the stag which dominates, usually with a magnificent set of antlers which would be the envy of the many sportsmen who now come from far and near to obtain such a trophy head. (The author can recollect the spring day when he found no less than 12 such sets of antlers cast by stags in a peat hagg at about 500m in Glenisla in Angus) It is the stag whose sporting value determines the value of a deer forest – in recent years this has risen to as much as £15,000 per animal, and may be one reason why today's landowners are unwilling to take the very drastic measures necessary to bring down deer numbers.

The related issue, which has become highly controversial, is the question of public access on sporting estates, made worse by the erection of notices discouraging access throughout the lengthy stalking season (if hind stalking is included). There is little doubt that an arduous stalk can be completely ruined by the sight of an orange anorak on the skyline, but equally, some owners and keepers are still loathe to accommodate the legitimate interests of the public and respect for the traditional 'right to roam' which has made the Scottish hills so accessible to generations of naturalists and walkers.

> And sweeter to my ear is the concert of the deer
>   In their roaring
> Than when Erin from her lyre, warmest strains of Celtic fire
>   May be pouring

Duncan Ban McIntyre 18th *century Gaelic poet and gamekeeper*

The Leaper

This is the name given to that 'king of fish' when it is not being referred to on fishing boats as 'the gentleman' out of superstition – a superstition which only attached itself to powerful symbols such as the Atlantic salmon, revered throughout the northern latitudes and celebrated in folklore of American Indians and Celts alike from ancient times, as a sign of the regeneration of life. More prosaically in Scotland the salmon is important commercially, both as a much sought-after sport fish and as an increasingly farmed product, exported world wide. And as described in Alexander MacDonald's 18th century Gaelic poem *Song to Summer*, 'speckled skin's brilliant hue lit with flashes of silver', the leaping salmon navigating waterfalls such as at the Falls of Shin in Sutherland, or even more mundanely passing up a fish ladder, has become a tourist attraction in a number of locations in Scotland, notably at Pitlochry. The whole mystery of salmon navigation across the Atlantic and its ability to find its home river excites the public imagination, fuelled by the spectacle of its courageous efforts to surmount river barriers during its spring and autumn runs.

Despite considerable declines in numbers in recent years on the great salmon rivers such as the Tweed, Spey and Tay, the latter has seen anything from 30,000 to over 100,000 fish caught in any one year, and the sport is still regarded as the premier freshwater fishing experience – with prices to match for rental or purchase of favoured stretches of water. The survival of the salmon faces a number of obstacles apart from natural physical barriers. The construction of hydro-electric schemes and water abstraction has certainly affected previously good salmon waters. Large-scale forestry has a major effect on river catchments and drainage as well as silting up spawning beds. Young salmon, and their eggs, have been killed by the effects of acid rain in some parts of Scotland. Compared to these influences, to say nothing of the huge effects of monofilament drift netting of the coast of the north-east of England, the impact of other predators such as seals and saw-billed ducks like mergansers and goosanders, must be relatively modest, despite claims to the contrary by fishermen and water bailiffs. A very real concern is the maintenance of the genetic integrity of the wild stock in the face of interbreeding with escaped or dumped fish reared in fish farms.

The Otter

With its playful habits, this protected species is one of the animals many visitors would most like to see, and not surprisingly, has been chosen as the symbol of the Scottish Wildlife Trust. The clean rivers and sea lochs of Scotland are the last stronghold of this mammal in Britain, but it still thrives in many parts of the Highlands and islands. The latest research by the Vincent Wildlife Trust shows that otter numbers have increased by more than 20 per cent in the last four years – a truly remarkable increase – so that the population is now at least 10,000. Most recently, it has been found in almost every district in Scotland, indicating a return to its former haunts. A considerable hunter, otters will eat about a kilogram of fish every day – at one time they were persecuted by gamekeepers because of this – and will also take crabs, frogs, and birds eggs, including those of domestic fowl if available. Usually about a metre in length, a third of this is taken up by its strong tail which often leaves a grooved track behind its webbed feet.

Otters are secretive and easily disturbed, so quietness and patience are necessary if you are going to be lucky enough to see one. This is most likely to be in the sea, fairly close to the shore, where they swim among the kelp (seaweed), or climbing onto rocks to feed. One of its more endearing habits is to enjoy mud slides into water, which sometime give away its presence in the locality, though more frequently it is indicated by its droppings or spraints left on stones marking its landward territory. Otters are commonly seen in the sheltered bays in Shetland, at Balranald on North Uist, and family parties are frequently seen from the A97 beside Loch Davan at Muir of Dinnet on Deeside. But your best chance of seeing this elusive creature is probably at the Kylerhea Otter Haven on the isle of Skye where the Otter Trail and special observation hide (with TV link) has been constructed by Forest Enterprise. One of the most surprising findings in recent times (see *Vulnerable Islands*) is that the native otters now appear to compete very successfully with invading North American mink by the simple expedient of killing them!

CHAPTER 5

Birds

SCOTLAND IS AN ORNITHOLOGIST'S paradise, while for the less expert visitor, birds have the considerable advantage of being generally much easier to see than the shyer mammals. Over 450 species of bird have been recorded in Scotland, with 175 of these regularly breeding: some 19 British birds only nest north of the border. In the notes below mention is made of a number of species (including several in topic boxes) which are of particular significance because of their rarity, the importance of their populations in Scotland (which for example, has most of the world's **Manx shearwaters**), their special protection or even re-introduction, or because, in the case of some wild geese, they pose a control problem. A number of species are officially declared in need of special protection measures, such as the **white-tailed sea eagle**, the **osprey, capercaillie, Scottish crossbill**, or the **red-necked phalarope**, whose entire British population is almost confined to the island of Fetlar in the Shetlands. In addition a recent report of the RSPB indicates that further efforts are needed to protect **black grouse, hen harrier, grey partridge and bittern**. Some species, such as the breeding **greenshank** are very much on the southern limits of their sub-arctic ranges and are restricted to limited habitats – in this case, the peatland bogs of Caithness and Sutherland, while others are commonly seen under quite different conditions including the wader species such as **oyster catcher** and **curlew** which breed in the glens and uplands and feed on the lowland estuaries.

Scottish crossbill

The wild mountain country is home to the magnificent **golden eagle**, with as many as 450 breeding pairs in Scotland, making it the stronghold for this species in Europe. As a result of a re-introduction programme based on the island of Rum, it has now been joined by the even more spectacular white-tailed sea eagle (see *The White-Tailed Sea Eagle*). Most visitors are now aware that the osprey can easily be seen through the special viewing facilities provided by the RSPB at Loch Garten. Perhaps not so well known is the regularly nesting pair protected by the Scottish Wildlife Trust at Loch of the Lowes in Perthshire. A visitor who witnesses the dramatic swoop of an osprey in the act of catching a trout in a great shower of silver spray is unlikely to forget the sight.

The rasping cry of the **corncrake** (a bird frequently heard in times past in cornfields throughout the country), can rarely be heard outside the Western and Northern Isles of Scotland – visitors may be lucky enough to hear it on the reserves of Balranald and Loch Druidibeg in the Uists. A rare breeding bird, the **Slavonian grebe**, with its extraordinary mating dance, can be found in small remote highland lochans, sometimes alongside those other very attractive sub-arctic birds, the **black-throated** and **red-throated diver** – both these and the grebes being extremely shy birds which are easily disturbed from their nests.

Red-throated diver

The native pinewoods of Scotland are home to some bird species which are found nowhere else. One of the most notable is the Scottish crossbill. It is the only bird which is unique to Scotland, very dependent on old Scots pine. Visitors who spot the capercaillie (see *Capercaillie or 'Horse of the Woods'*) for the first time may be forgiven for thinking they have seen a wild turkey, since this is the largest of the grouse family, and can be quite an intimidating bird when defending its territory. By contrast, the crested tit is minute and is more likely to be heard than seen with its repeated high thin call in the upper canopy of the pines of Speyside. The **ptarmigan** in Scotland is entirely confined to the Highlands, and is a truly arctic grouse which can survive

The White-Tailed Sea Eagle
– A Successful Re-introduction

Previously nesting throughout much of our rocky northern and western coasts, this magnificent bird, with a wingspan of over 2m, and the largest bird of prey in Britain, was persecuted to extinction in Scotland at the beginning of this century, despite its known 100 breeding sites here. The last native pair of sea eagles bred on Skye in 1916. This great 'eagle with the sunlit eye' as the Gaelic has it, is known for its fantastic courtship display when a pair in mid air will cartwheel earthwards with their talons locked! Its diet consists mainly of fish, seabirds, rabbits and hares and some carrion in the winter months, though there have been reports of sheep being taken on Mull. It was successfully introduced from a base on the National Nature Reserve of Rum in the mid-1970s in a joint programme between the RSPB and SNH. Requiring some 5 years to reach breeding maturity, there was an anxious wait for the first eaglets brought over from Norway to show signs of nesting in 1982 and it was 1985 before the first young bird bred in Scotland soared over the west coast. (Eleven more young sea eagles were brought from Norway in 1997). Twelve pairs of sea eagles nested in 1997, with 5 pairs rearing 9 young, to bring the total to over 60 fledged chicks - a genuine conservation triumph.

under the most extreme conditions at high altitudes – one of the hardiest birds in the world – characteristically white in winter and mottled brown and grey at other times.

The coastal cliffs and islands of Scotland support some of the largest populations of breeding seabirds anywhere in the world – perhaps as many as 6 million birds during the breeding season, including more than half the world's population of **gannets** (see *The Solan Goose*) with their spectacular plunging dive. The biggest gannetry in the world is on St. Kilda, but much more accessible is the Bass Rock in the Firth of Forth, barely an hour from Edinburgh. That colourful comic of the grassy cliff tops, the **puffin,** can be seen on the Isle of May (among a number of other sites) and which also has a variety of other seabirds, including thriving colonies of **shags, kittiwakes and guillemots.** These and many other bird islands can be reached by regular boat trips, but

Shag

most of the species can also be easily found in considerable numbers on the mainland coast, such as at St. Abbs in Berwickshire, Seaton Cliffs in Angus, Fowlsheugh near Stonehaven and on the Buchan Coast – at Longhaven around 23,000 birds of nine species nest on the cliffs. The visitor will usually have to travel to Caithness or to the Northern Isles however to be dive-bombed by the piratical bonxie or **great skua** (see *Don't Get Skewered by a Skua*) – a unique wildlife experience!

The machair of the Western Isles are notable for the sheer density of breeding waders and other ground-nesting birds, including oyster catcher, **lapwing, dunlin, redshank** and **ringed plover,** whose calls fill the air in late spring. The RSPB reserve at Balranald in North Uist is outstanding, with almost a thousand nesting pairs of different species. The great wet peat flows of Caithness and Sutherland are also an ornithologist's mecca, with their nesting greenshank, **golden plover** and curlew, in addition to the breeding black-throated divers on the many small lochans in this district. This internationally important bird habitat can best be seen at the RSPB reserve at Forsinard or from the nearby A897.

During the period between autumn and spring many of these waders can be viewed along the shores of the estuaries, together with huge flocks of overwintering duck and geese. On Islay, especially round Loch Gruinart, and at Caerlaverock on the Solway, large numbers of **barnacle geese** roost on the sandbanks and feed on the nearby fields. The lochs and lowlands of Angus and Fire provide a roost for great flocks of **greylag geese,** now numbering over 10,000, and smaller numbers of **pink-footed geese** – some

Golden plover

Scottish Bird Names

Although many of the older Scots names for birds are little used nowadays, except perhaps locally, there are still such common ones as *whaup*, for curlew, *mavis* for thrush, *hoolet* for owl, and *peesie* for lapwing or peewit. Many words characterise the bird's call, behaviour or appearance – such as *bletherin tam* for whitethroat, *willie-whip-the-wind* for kestrel, *blue bonnet* for blue tit, and *bull-o'-the-bog* for bittern. *Gowk* for cuckoo now also describes a 'daft' person, (a gowk storm of sudden spring snow showers often comes with the first cuckoo), and *waggitie* for the pied wagtail is obvious. Several are simply a delight to listen to like *wallopieweet* for lapwing, *feltieflier* for fieldfare, and *tibbie theifie* for sandpiper. There is affection for the puffin in *tammie norie* or *tammie cheekie*, just as there is perhaps irritation for the noisy *skirly wheeter* or oyster catcher or the incessant sighing in the tree tops of the *cushat doo* or wood pigeon.

23,000 of these arrive at Loch Leven each autumn. The flighting of these vast congregations of birds, especially as they come into their evening roosts, is a most memorable sight and sound. Over 20,000 **eider** duck collect in the Tay Estuary during these months, and the Firth of Forth, notably in the sheltered bays such as Skinflats and Aberlady Bay on the southern shore, holds some of the most important populations of a wide variety of diving duck, including **golden eye, long-tailed duck, pochard** and **shelduck**, as well as nearly 1,000 **crested grebes**.

Eider

Chasing the Wild Geese

Scotland's estuaries, islands and grasslands are key resources for wintering wildfowl migrating from their arctic breeding grounds to feed here before their spring return north. The insulation provided by our island situation combined with the warming influence of the Gulf Stream means that, in contrast to other potential wintering grounds at similar latitudes, Scotland's estuaries are very rarely frozen for extended periods – the last significant freeze-up was in 1963. Scotland is therefore important internationally in holding over the winter months either whole populations, or very high proportions of the total population, of particular species such as, the Sptizbergen and Greenland races of barnacle geese or the **Greenland white-fronted goose**, which has a relatively small world population.

These great flocks normally feed on natural marshland (known as *merse* in Scotland) but now increasingly, on coastal farmland, with the trend towards winter cereal growing and early grasses. After feeding during the daylight hours, the geese move on to safe roosts, either on intertidal mudflats or suitable inland waters. The silhouette of literally thousands of geese dropping down to a roost against the pink sunset of a winter evening and filling the air with their honking, is a truly memorable one. Unfortunately it can be more than memorable for the farmers whose land is used for feeding – it can be an expensive luxury. While at one time farmers were generally tolerant of their winter visitors at acceptable levels, there is no doubt that in certain localities, for example on the Solway and on Islay, the number of birds has risen significantly in recent years. This may be partly as a result of better protective legislation, including a ban on the sale of wild geese, and partly to a series of successful breeding seasons.

The problem is compounded by changes in agricultural practice: whereas in the past, farmers might have little concern about birds cleaning up potato fields after harvesting, or feeding on stubbles, they are understandably less tolerant of feeding on winter cereals or the 'early bite' of grass denied to sheep. Goose damage is not confined to the removal of green material (which some claim may even be advantageous in stimulating growth) but also results in soil compaction, puddling, and extensive fouling by goose droppings.

The problem has been acknowledged in the system whereby farmers are entitled to apply for licences to shoot geese which can be shown to be causing serious damage. Within statutory Sites of Special Scientific Interest (SSSI), they can be compensated for damage through management agreements – now costing tens of thousands of pounds each year. Scaring geese by mechanical means is rarely successful as over time, the geese adjust to the regular bangs. The alternative of using 'human scarecrows' to chase the geese has been tried in recent years with some success – hence the title above. However farmers can claim with some justification that those outside SSSI are discriminated against (even if they have recourse to shooting licences where damage can be shown) and there have been calls for more general culls to bring down numbers overall.

In the case of specially protected species such as barnacle and Greenland white-fronts, this would require a special dispensation under the international directives which require governments to safeguard such species. Conservationists argue that the volatility of breeding success is such that any such general cull would endanger vulnerable populations. 'Chasing the wild geese' in Scotland represents a classic conflict between long-

Shelduck

Flying a Kite

The red kite is the most recent successful bird re-introduction to Scotland, with a record number of young flown from the nest in 1997 – at least 25 pairs attempted to nest, of which 19 pairs raised 40 young – an increase on the previous year. The first successful breeding took place in 1992. Two hundred years ago, this bird would have been relatively common throughout Scotland, but the last native kites probably disappeared around 1876, partly as a result of persecution, and partly because of the removal of carrion from towns over the centuries. Kites look superficially similar to buzzards, but are more slender, have longer angular wings (about 2m) and a very distinctive forked tail. Highly efficient scavengers and feeding mainly on dead rabbits, crows, sheep and other carrion, they seem to pose no threat to farming or sporting interests. Re-introduction to northern Scotland from Sweden commenced in 1987, and has now been extended to central Scotland in a project administered by the RSPB and with the support of Scottish Natural Heritage. Tail-mounted radio transmitters have enabled the movements of some of these birds to be tracked – while some have headed across the Irish Sea to County Antrim, one wise individual apparently travelled the 400 miles to winter in the more benign climate of Paignton in Devon!

Red kite

term international conservation objectives and local agricultural economic interests which has yet to be resolved satisfactorily.

Capercaillie or 'Horse of the Woods'

One of the names for this largest member of the grouse family, the 'Horse of the Woods' may indicate its size and clumsiness, for it is a noisy flier and when disturbed in its usual habitat, usually conifer forest, it crashes through the foliage and branches. A diagnostic sign of the presence of capercaillie is its droppings on the forest floor, largely comprising pine needles. The call of the male is equally distinctive, being a series of harsh clicks becoming increasingly rapid, and ending with a sound reminiscent of a cork being pulled from a bottle. A favourite target for sportsmen, it became extinct in Scotland in 1785 with the shooting of the last two birds in Aberdeenshire in that year. It was re-introduced successfully into Perthshire in 1837 with birds brought from Sweden. Now it is again in serious decline for reasons which are not at all clear, perhaps due to loss of habitat, but cold wet spring weather can also be lethal to the young birds. What is now known however is that there is significant mortality from birds flying into deer fences (from the number of birds found dead along these lines) and that this approach to forest protection has serious consequences for this and other bird species.

Like other members of the grouse family, the 'caper' has aggressive pre-mating displays when establishing or defending its territory and will even attack humans – this turkey-sized bird, fanning its tail and running at speed with head raised and large beak open, is an intimidating sight! Although often associated with native pine-woods, it can survive happily in other conifer plantations – at one time it was regarded as a forestry pest because of its diet of conifer shoots and buds, and was persecuted accordingly. Although classified as a game bird, there is now agreement between a number of estates that to restore its numbers, shooting of the capercaillie should be at least temporarily halted.

Don't Get Skewered by a Skua!

Both the great and the **arctic skuas**, unique among Scottish birds (with the possible exception of nesting lapwings which can put a dog to flight) have the less than endearing habit of dive-bombing intruders into their breeding territories, often with an aggression which might be described as coura-geous! Their skimming flight from any angle soaring upwards only at the last moment can be quite intimidating and the author has found it useful on occasions to carry a walking stick at head height if only to give a sense of security. Of the two, the smaller arctic skua, with its elegant and more agile flight, and extended tail feathers, often attacking silently from the rear, is the more frightening. The great skua, or *bonxie*, almost as big as a **buzzard**, with conspicuous white wing patches, has been known to kill birds as large as gannets and will at least give warning of its attacks with its grunting call. Both species harass other birds such as kittiwakes and **terns** for their fish catches.

However, the most astonishing figures, as a result of recent research, are the numbers of other seabirds actually killed by great skuas – up to 200,000 in Shetland in one recent year, while whole colonies of kitti-wakes have apparently been wiped out. It has been estimated that the 400 bonxies on Noss (see *Places to Enjoy Wildlife*) ate more than 4,000 kit-tiwake chicks in 1997. There have been suggestions that this predation is linked to the dramatic fall in sand eels in the 1980s and the recent depar-ture of 'klondykers' – large foreign vessels which take the catches from local fishermen – and who discarded considerable quantities of fish which the bonxies exploited. In global terms, because of its restricted popula-tion, the great skua is undoubtedly Shetland's most important seabird, so there is now a considerable dilemma regarding the balance to be struck between protecting this species and maintaining the diversity of other wild birds on the islands.

Something to Grouse About

At the other end of the size scale from the redoubtable capercaillie, that favourite target of the sportsman, the red grouse has been in decline for the last 50 years or more.The male has a distinctive red comb and a call which can hardly be mistaken "go-bak, go-bak" as its explosive take-off from the heather ends in a gliding curved- wing flight some distance away, accompanied by its whirring wing beats. It is quite dependent on extensive heather moorland and the young need a good supply of fresh shoots of this plant, usually maintained by periodic burning. Unfortunately the grouse really *has* something to grouse about, in the reduction of its habitat by increasing forestry plantations and encroachment by grassland, usually as a result of increase in sheep grazing. To add to its woes, the results of surveys along deer fence lines in the Cairngorms has shown that of the 437 bird collisions recorded, the highest percentage (42 per cent) were of red grouse, the remaining being shared between black grouse and capercaillie, with about half of these collisions resulting in death or severe injury. Together with the ravages of the heather beetle, disease, and locally, an increase in predation, and in some seasons cold wet spring weather which is lethal to young birds, numbers dropped to such an extent on some sporting estates that the traditional 12 August start to the shooting season has had to be abandoned. However, 1997 saw a remarkable change with good numbers of birds on many estates.

Red grouse

The Corncrake – A Recovery ?

Once widespread throughout the lowlands of Britain and now a globally-threatened species, the corncrake has declined drastically since the days when Lord Cockburn in the early 19th century listened to its distinctive call in Queen's Street Gardens in central Edinburgh! Corncrakes favour in particular a combination of weedy arable land, rough vegetation and hay meadows, all in short supply in today's intensively farmed countryside. With its chemicals and early cutting of hay for silage, modern intensive agriculture has undoubtedly forced the retreat of this shy bird to its last strongholds, mainly in the western and northern isles of Scotland, usually under crofting systems where traditional agriculture is still practised. For nesting, it depends on deep cover, such as those provided by rushes and irises, which are still common in the islands.

In 1993, the total population was estimated at 480 singing males, a decline of 17 per cent over a mere 5 years. This prompted the RSPB, SNH and the Crofters Union to collaborate in a management programme aimed at maintaining the 1993 population or even better, increasing the number and distribution of the species, which at that time was restricted to 82 kilometer squares. By 1996, the numbers of calling males had risen to 637 in core areas, though there are considerable unexplained fluctuations from year to year. However, there are now good prospects for the survival of the species, at least in the northern and western districts, where the rasping call of this elusive bird is so characteristic of Hebridean summer nights. The island of Coll in the Inner Hebrides (see *Places to Enjoy Wildlife*) is a good place to hear this evocative bird.

Harrying the Harriers

Birds of prey, much beloved by ornithologists, have nevertheless in recent years aroused the ire of a number of quite different interests, for example, pigeon fanciers who have accused **peregrine falcons** of devastating their sport by the number of kills of highly-prized and very expensive racing pigeons. The same species share the moors with another bird which has come in for particular criticism from grouse moor owners and gamekeepers – the **hen harrier**. Although this fine bird is now widely scattered throughout Scotland, before the 1940s it was largely confined to the moorlands and grasslands of Orkney where it preyed extensively on the many pipits and Orkney voles present. Its remarkable increase on the mainland has been helped, partly by the decline in gamekeeper activity, but also by the increase in young conifer plantations which, protected from sheep, produce in their early years the long grass cover which voles favour. Although it can at a distance be mistaken for a large gull, the harrier has a most acrobatic flight, sometimes suddenly sailing upwards, then rolling over and diving, often quite close to the ground in what seems to be a random pattern, with the wings forming a shallow V.

Harriers feed on a wide variety of rodents and birds including grouse, but their numbers can increase substantially where sheep grazing has increased the proportion of grass within the moorlands, usually below 460m where pipits are present and more often on eastern than on western highland moors. They will undoubtedly take young grouse where available and also stand accused by keepers of scaring adult birds during shoots. Notwithstanding claims that they have vastly increased in number in most recent years, expert opinion is that, while there may be local rises in populations, there has in fact been a general reduction in numbers over the last 10 years, so that they are hardly more numerous than the relatively scarce golden eagle. Nevertheless, there have been powerful calls by shooting interests to have these protected birds of prey culled.

The most recent experimental studies on the hen harrier 'problem' has proved inconclusive. On one estate where the birds were rigorously protected, the number of harriers increased dramatically over 5 years, and some of the highest densities of predatory birds ever were recorded, indicating quite clearly the effect of keepering combined with control of

Preying on Birds of Prey

It is an unfortunate fact that persecution of birds of prey is still common in Scotland. In 1997, the RSPB reported 141 alleged offences, 92 of these claimed to be direct persecution and 49 the illegal use of poisons. The 21 poisoning incidents confirmed involved buzzards, golden eagle and white-tailed eagle, the last two species being rare. Birds killed by means other than poisoning, such as trapping and shooting, included buzzards, hen harriers, peregrine falcon, **kestrel**, **sparrowhawk** and **long-eared owl**. Almost all of the incidents take place on managed upland grouse moors or on low ground pheasant rearing areas. Eggs are still taken by collectors, including ospreys' and eagles' eggs. Although not usually thought of as birds of prey, fish-eating birds such as **cormorants, goosanders** and **mergansers** and even **herons** are increasingly becoming the target of sport fishing interests. There is good evidence to show that the publicity given to such cases and the increasing fines (the record to date is a fine of £90,000 imposed in 1997) is making the perpetrators more wary and careful in their activities to avoid detection, which is always extremely difficult.

such other predators as crows and foxes. Nor was there any doubt about the drop in the numbers of grouse shot by the end of this period – but equally, the general decline in grouse numbers throughout Scotland cannot be attributed to predation by birds of prey, which in the case of the study area were quite artificially high. Given the extent of land in Scotland dedicated to this form of land use, the issue remains an important one both for conservation and the economy of sport shooting.

The Solan Goose

The old name for the gannet may relate to its usefulness as a food item – young birds known as gugas are still collected from the remote island of North Rona in the Atlantic by the men of Ness at the northernmost point of Lewis as part of a centuries-old tradition. Colonies are known from as far south as the Scare Rocks in the outer Solway to St. Kilda, a hundred miles out into the Atlantic. Before the evacuation of St.Kilda, gannets and their eggs and feathers were an important resource for that community. There are other interesting records of gannets as a source of food, for example, the 16th century soldiers garrisoned on the Bass Rock were expected to exist on the fish carried to the rock by the gannets, and to keep warm by burning the material the birds had collected for their nests. One method of capturing the birds for food was to attach herrings to a board, so that when they dived with their usual velocity, their sharp beaks became embedded in the wood. Gannets were at one time sold in Edinburgh for two shillings each.

Recent surveys have confirmed Scotland as a stronghold of this species, with 12 colonies and a total population of nearly 170,000 occupied nests, representing over 61 per cent of the eastern Atlantic population. St. Kilda alone has over 60,000 birds, and the gannetries around the coast of Scotland make up a high proportion of the total world population, which takes its Latin name, *Sula bassana* from one of its most famous sites, the Bass Rock (see *Places to Enjoy Wildlife*) in the Firth of Forth. A very handsome bird with a pale cream coloured head, its black wing tips heighten the snowy-whiteness of its body plumage. The head and neck contain air-filled cells which cushion the shock of its high speed dive when fishing sometimes from 30m or more – a dive made all the more spectacular by the change from an apparently lazy casual flight into a twisting torpedo plummet with folded wings. While the sight of a small group of gannets hunting along the coast is fascinating enough, there is nothing to compare with the experience of being in a boat near a colony with thousands of these superb sea hunters plunging all around, and watching their elegant ghostly shapes swimming smoothly through the green waters below.

CHAPTER 6

Insects

OF THE 14,000 INSECTS in Scotland, 1,300 occur nowhere else in Britain, and three are endemic. The most important habitats are pinewoods, birchwoods and herb-rich moorland. Many of the species are highly restricted in their habitat. The **Rannoch sprawler** moth occurs in very old birch trees in Speyside and Rannoch, while another species is restricted to the same tree species in Glen Affric. The Scottish insect fauna includes many northern species at the edge of their range, including the majority of the butterflies and moths, which are the most conspicuous of our insects. Compared to the rest of Britain, Scotland is relatively poor in the number of resident butterflies, with only 28 species, including several arctic-alpine species.

A good example is the **small mountain ringlet,** found at altitudes up to 1,000m in Perthshire, Argyll and Invernesshire – a good place to see this butterfly is on the National Nature Reserve at Ben Lawers (see description under *Places to Enjoy Wildlife*) Remarkably similar in appearance is the **Scotch argus** which is widely distributed on moorlands in Scotland, but only occurs in England at two localities. Clearly distinguishable by its red or orange (dependent on sex) eye-spotted bands on its dark brown wings, this butterfly is often found close to woodlands or wet places from July to September.

Wood ant

Two species of butterfly in Scotland have been identified as in need of special protection and management. These are the **pearl bordered fritillary** and the **marsh fritillary.** The latter is in now very much threatened in Europe, and Scotland remains one of its last strongholds. Like so many other lepidoptera, this very attractive small brown and orange butterfly is quite specific in its food plants, and the larvae feed exclusively on **devil's bit scabious,** which itself is restricted to hill pasture, moorland or damp meadows. For this reason, it is now confined almost entirely to Argyll and parts of Inverness-shire and Dumbartonshire, apart from some of the Inner Isles where its sedentary habits inhibit its expansion.

A butterfly considered to be endemic to Scotland is the **northern brown argus** (or Scotch brown argus), identifiable by its very dark brown wings with a distinct white spot on the forewing and bright orange patches along the hind margin of the rear wings. Like the marsh fritillary and

---

Antics in the Woods

Wood ants nests are relatively common in the pinewoods and birchwoods of northern Scotland, recognisable as symmetrical heaps, usually up to a metre high, of what initially looks like a large pile of brown pine needles. These nest mounds are built either by the **Scottish wood ant**, a nationally scarce species, or the **hairy wood ant**, although in Glenmore, the nests of the very localised **narrow-headed wood ant** can also be found. They collect heat as a result of warming by the sun, in addition to that supplied by the sunbathing worker ants, which then carry their heated bodies into the nest chambers of the brood and the queens. The nests are always positioned within the woodland to receive the right amount of sun in the appropriate season - in itself a remarkable phenomenon.

One of the fascinating aspects of their behaviour is their capacity to milk aphids of their honey dew, but they also prey on a wide range of insects and other invertebrates. Well-worn trails from the nest to favoured trees for milking aphids and predating caterpillars are a feature of the woodlands inhabited by these ants. Large networks of colonies are created by the linking of other nests in the neighbourhood, with new nests often formed by budding from existing nests, while worker ants are sometimes exchanged between nests. Visitors should not disturb the nests of these beneficial insects, which are threatened by forest removal and overgrazing.

---

many others, the larvae are dependent on a single food plant, in this case, the rock rose. It is usually found on well-drained grassy slopes on sites scattered across much of Scotland, as, on the pastures behind the cliffs at St. Abb's Head in Berwickshire (see *Places to Enjoy Wildlife*) but it was known in the 18th and 19th century on Arthur's Seat in Edinburgh's Holyrood Park, before it was collected to extinction by dealers. The species, which is very thinly distributed, is vulnerable to changes in grassland management, including overgrazing.

Set against these declining species, there is the success story of the **orange tip**, which until the 1950s appeared to be confined to two areas in the Borders and the north-east, although 100 years prior to that, it was common throughout Scotland. With its bright orange patch tipped with black on the outer edge of its forewing, this species is unmistakable. Unlike the northern brown argus, which appears to depend on a parasite to control its population, the larvae of the orange tip are cannibalistic, so that only one survives on any one plant, thus ensuring its food supply. Now the orange tip is widespread across central Scotland and seems to be spreading northwards, so that it is probably commoner than ever before here.

Emperor moth

The day-flying **burnet moths** are especially distinctive among the family of butterflies and moths in having shining black wings, decorated with variable numbers of bright red spots and equally distinctive black club-shaped antennae. Several of the burnet moths of Scotland are of special ecological interest because of their peculiar distribution and genetic character – their origins are still matters of speculation. In Scotland, several common species are found on coastal grassland or upland moorland, three of which in Britain are restricted to Scotland. Some of the species are very rare or sparsely distributed, and the **New Forest burnet moth** is restricted to a solitary site in Argyll in western Scotland, where special protective management is undertaken. Another, the **slender Scotch burnet**, is confined to the grasslands associated with basalt lavas on Mull. The only species of burnet moth found in the uplands, the **mountain burnet**, which feeds on crowberry, is confined to two or three hills near Braemar in Aberdeenshire, distinguished by its hairy body and five relatively dull red spots.

The most widespread of the group, the **six-spot burnet**, is extremely abundant in grassy hollows in dunes, often feeding on the nectar of rose bay willow herb. Frequently mistaken for a butterfly, the moth's caterpillar is creamy yellow, with a series of black spots along the sides and back. Like other burnet moths, if crushed, it gives off cyanide dangerous to any predator. Other than the mountain species, most burnets prefer warm grassy hollows where there are plentiful vetches, their main food.

Six-spot burnet

CHAPTER 7

The Marine World

WHILE SCOTLAND'S TERRESTRIAL WILDLIFE riches have been acknowledged for some time, it is only relatively recently that we have begun to fully appreciate the wealth of our marine heritage and its potential contribution to visitors' experience. The North Atlantic Drift produces a mix of up-welling warm and cold waters which helps to make our coastal waters among the richest in the world – in the vicinity of St. Kilda (see An Underwater Eden) this has created a marine environment of extraordinary diversity. It is reckoned that a bucketful of sea water from this area would contain around 4 million tiny plants, animals and bacteria. The marine world around our coasts is thought to contain 7,000 species of invertebrates and 900 species of plants, with organisms ranging from **basking shark**, a regular visitor to Scottish waters (the second largest fish in the world, up to 14m in length and weighing up to six tonnes), to the microscopic **plankton** which it feeds on, while few people realise that Scotland has in the so-called *maerl* beds its own extensive living and fossil coral beds.

With a coastline of over 12,000km, including many islands and rock skerries, and deep narrow sea lochs, Scotland has an unusually diverse sequence of landforms and wildlife where the land and the sea meet, with everything from windswept cliffs, among the highest in Europe, to great stretches of shell sand. The **kelp** forest at this meeting point is an especially important refuge for many plants and animals.

At certain localities, visibility can go down to 36m with spectacular underwater cliffs reaching down to depths of over 100m. Many of the rocks are covered in colourful organisms, including **cup corals, dead man's fingers, sea anemones, hydroids, bryozoans, chitons** and many other molluscs. On the west coast, Loch Creran has the very rare marine worm *Serpula* which lives inside a hard chalky tube, but here is almost unique in forming dense colonies which also host **sponges, brittlestars, starfish** and **sea quirts**.

This marine environment is of considerable economic importance for both its finfish and shellfish, producing **cod** up to 1.5m long and weighing up to 30kg, or **herring** shoals which can number about 500 million fish per square mile. Other species such as **plaice, mackerel** and **sole** are becoming of increasing importance, while the price for Scottish shellfish such as **prawns, lobsters, crabs** and **scallops** is an indication of their recognised quality. The sea bed is full of the burrowing **Norway lobster,** often known as scampi, which is probably the most important shellfish of all. **Sand eels** provide food for many other species, not least many of the great seabird populations which are a feature of the Scottish coast. Of all the fish landed in Britain, over 70 per cent is landed in Scotland, with Peterhead now the largest fishing port in Europe.

There is increasing evidence of warming of our seas from the recent catches of distinctly southern species such as **sun fish, garfish** or **launce, trigger fish** and large numbers of the deep blue jellyfish known as **Jack sail-by-the-wind** and **lion's mane jellyfish** up to 50cm in diameter. (The sudden appearance of much smaller jellyfish in huge numbers were responsible in 1997 for the deaths of many farmed fish by clogging up their gills). It may that warming which was responsible for the finding, early in 1998 on the Aberdeenshire coast, of one of the biggest, but most reclusive, sea creatures – **the giant squid.** Capable of growing to a length of almost 20m, this is a creature so hard to find that it is only within the last 130 years that scientists have been able to prove that it really exists.

Among the most appealing of our marine mammals, **seals** are found in large numbers round the whole of the coast of Scotland. Supremely elegant when swimming underwater, they are skilled hunters of fish and other prey. They can dive for up to an hour to depths of over 200m and even down to 500m without coming up for breath. In addition to the two main species, others such as **bearded seals, harp seals, ringed seals** and **hooded seals** also occur. The two species which breed in Scotland are the **grey seal** and the **common seal.** Although both spend most of their time at sea, they come ashore to breed and moult, and can be seen on isolated rocks and skerries offshore, apparently sun-bathing and often arching their backs so that they appear saucer-shaped. The author can vouch for the fact that they have a sense of humour as he watched one off St Abbs Head in Berwickshire deliberately surfacing from time to time with a fish in its mouth to tease a very frustrated circle of hungry gulls attempting to wrest the fish away! They are known as curious creatures, responding to music and frequently swimming up to boats to see what is going on.

Grey seals are distinguished by the 'Roman' nose of the male contrasting with the more dog like face of the common seal. Common seals tend also to be more mottled – they can be very different between individuals and between different regions in colour

Grey seal

and pattern – while grey seals are distinctly larger when mature, the males growing up to 2.3m, compared with 1.5m for the male of the common seal. The common seal does not form as large groups as the grey seal. The largest living carnivore in Scotland, it is estimated that more than 90 per cent (over 100,000) of the UK population breed in Scotland, with pup production increasing by 7 per cent each year. There are thought to be at least 27,000 common seals in Scotland. The grey seal tends to favour the rockier western coasts, while the common seal is more often seen in the sandy estuaries and firths.

The most important breeding sites for both species of seals are in the Northern and Western Isles and parts of the west mainland coast. The largest numbers are to be found in Orkney, but the Monach Isles off Uist have the second largest breeding colony of grey seals in the world. Some of the biggest breeding populations are to be found on the remote islands of St. Kilda and North Rona. Both species of seals are commonly found together. The best times to see seals at their breeding sites are from mid June for common seals, with moulting taking place in August and September. Grey seals arrive at their sites from the middle of September, with moulting in late winter and early spring. Whereas grey seals mate on land, common seals mate in the water. Of course seals can be seen almost anywhere both in the water and out throughout the year – one of my favourite views is watching them from the railway bridge, basking on the sandbanks of the River Tay. They can be guaranteed to be seen in Lochs Dunvegan and Bracadale at the north end of Skye, and there is a notable breeding ground of common seals on the Crowlin Islands south-east of Raasay. Grey seals can be found around the Isla of May in the Firth of Forth. Further inland, seals can be seen in the sea lochs of Loch Laxford, Loch Etive and occasionally Loch Ness and they can regularly be seen between Dingwall and Evanton on the Cromarty Firth.

An Underwater Eden

Biologists using a mini-submarine in the rich waters around St. Kilda have recently made some remarkable discoveries about the marine life in this part of the North Atlantic. Described as 'an underwater Eden' the scientists have found a world teeming with wildlife contrasting starkly with the remote barrenness of the island landscape above. They have described seeing every square inch of rock covered with sea life, from a new species of sea squirt to spider crabs and delicate jewel anemone inside huge forests of kelp. Vertical sea cliffs plunge 50m to the ocean floor, which itself had boulders larger than two-storey houses. Large unknown sea caves have been found, providing yet another exotic habitat for specialised marine life. More than 400 different species have been discovered, including 45 kinds of molluscs, 20 types of sea squirt and 23 different species of fish from the prosaic herring to more exotic leopard-spotted goby. The bands of kelp stretched to 3 times the depth normally found in the Western Isles, due to the clarity of the water, supporting many animals which would not otherwise be found at these depths. All of this re-enforces St. Kilda's status (see *Places to Enjoy Wildlife*) as the only natural World Heritage Site in Scotland.

Some of the richest marine life is to be found off the Hebrides, which not so long ago supported a whaling industry, based on many different whale species – **blue whale, sei whale, fin whale** and **humpbacks** – were regularly caught. In the early part of this century, Scotland had no less than 6 whaling stations, including four in Shetland, while the last one in Harris only ceased operations in 1952. The area is now visited regularly by whales, **dolphins** and **porpoises**, a group collectively known as cetaceans. The largest animal known to have lived on earth is the blue whale at more than 30m long weighing 120 tonnes. By comparison, some of the smallest dolphins and porpoises are less than 2m long.

About 26 species of cetaceans have been recorded in Scottish waters, fifteen of them regularly. **Killer whales** are sometimes seen, travelling around in family groups known as pods, although they rarely remain in one area for more than a few days. They are easily distinguished by their tall dorsal fins. (Killer whales have been implicated in the recent finding of headless seals washed ashore in East Lothian, where sightings of this species have increased in recent years.) It is considered the longest-lived of any cetacean, with males living for up to 50 years, but females sometimes achieve the ripe old age of 90! The **minke whale**, one of the smaller animals in the group known as baleen or toothless whales (including the blue whale referred to above) was known as the 'herring hog' from its capacity to gulp down a whole shoal of that fish at one time. There is a seasonally resident population of this species around the waters of Mull, where individual whales have been recognised year after year. Minke whales are more common here than anywhere else in Britain. There is evidence of a recent increase around Scotland of humpback

---

Marine Ram Raiders

The image of the friendly dolphin with a sense of humour has taken a beating recently with authenticated sightings of dolphin attacks on porpoises, from the Moray Firth to the Forth. Post mortems carried out on porpoises in the Moray Firth over a 3year period showed that well over half, amounting to 42 (mainly juvenile) porpoises, were killed by **bottle-nose dolphins**. Most of the animals apparently died from high-speed ramming, causing serious internal injuries. In 1997 passengers on a pleasure cruise boat in the Firth of Forth witnessed a bottle-nose dolphin repeatedly butting a porpoise and throwing it into the air. Scientists still have no idea why these attacks occur. Dolphins themselves can be harassed by those dolphin-watching boats which insist on following too closely, and in the Moray Firth, where cetaceans have become an increasingly popular tourist attraction, a code of conduct has now been published to minimise disturbance to these marine mammals. Labelled the 'Dolphin Space Programme', it has now been agreed by most of the wildlife cruise operators in the area.

whales, which have been seen off Skye and even the Clyde. Schools of over a hundred porpoises have been recorded in these western waters.

Bottlenose dolphins such as those regularly observed in the Moray Firth usually live in groups of up to four individuals, although herds of twenty or more animals can be seen in the autumn. Cetaceans can be found almost anywhere around the coast of Scotland, especially in the islands and in the north and west. One of the most popular places in Britain is the Moray Firth with its unusual resident population of bottlenose dolphins, and **harbour porpoises** in small groups – both are frequently seen from the shore. Other good shore sites occur from Banff in the east to Balintore in the west, especially from Chanonry Point and North Kessock – because of the risk of disturbance to cetaceans, visitors are especially encouraged to watch them from land. **Common, white-beaked** and **Risso's dolphins** are commonly seen in the course of whale-watching boat trips off the west coast, for example, those operated by Sea Life Surveys on the west coast of Mull. Deep Sea World at North Queensferry near Edinburgh and Sea Life Centres near Oban and at St. Andrews provide a useful introduction to marine life in Scotland.

Minke whale

With the increasing number of small boats available to take visitors out specifically for this purpose, cetaceans are becoming much easier to see around the Scottish coast. Apart from the Moray Firth, both bottlenose dolphins and harbour porpoises can also be seen in the Cromarty and Dornoch Firths – the latter is the most important breeding location for seals, but outside the breeding season, some 1,200 common seals and 500 grey seals can be seen in this area. Minke whales are commonly observed from the shore as are killer whales.

One of the surprises of recent years has been the realisation of how relatively common marine turtles are in Scottish waters, especially the **leatherback turtle**. The biggest species of turtle in the world, weighing up to 916kg and measuring up to 2.9m it is the most common of the turtles to visit the Scottish coast in the summer months. Although marine turtles breed on tropical and sub-tropical beaches, they migrate to feed, and are frequently found in temperate waters. Because leatherback turtles are apparently warm-blooded (unlike most other turtles) they can adjust to the relatively cold temperatures of Scottish waters and continue to feed on their main prey of jellyfish. (Unfortunately they can also confuse the latter with discarded polythene bags and suffocate to death) Leatherbacks can be recognised not only by their size, but also because they have a semi-rigid leathery and ridged carapace, or shell, compared to the usual hard-plated outer shell of all other marine turtles. Of the 171 records of marine turtles in Scotland in recent years, half of these have been seen since 1980. Marine turtles are globally threatened, being used for food, and decorative articles of leather and shell, as well as being caught frequently in fishing nets.

CHAPTER 8

Conserving Natural Scotland

NATURE CONSERVATION IN SCOTLAND is achieved in 3 main ways:

* *By protection of special areas such as reserves, scenic areas, etc.* (many of which are briefly described under *Places to Enjoy Wildlife*)

* *By general and specific conservation of a wide range of species, several of which are protected under international law and for which there are individual Species Action Programmes* – a number of these are highlighted in Topic Boxes

* *By the advice and encouragement provided to many agencies and individuals, including grant aid to protect the countryside, its habitats, wildlife and landscapes – for example the incentives offered to farmers and others in Environmentally Sensitive Areas*

All of these are underpinned by many different interlocking systems of environmental protection which are important in providing general frameworks for the regulation of changes in land use. These include the development permissions required under Town and Country Planning regulations administered by local authorities within the context of Structure and Local Plans, and the controls exercised on emissions into air and water by the Scottish Environmental Protection Agency. These and the increasing duties of government bodies such as the Forestry Commission and agricultural agencies provide the general setting for the work of such official organisations as Scottish Natural Heritage (SNH) and the several voluntary conservation agencies such as the Royal Society for the Protection of Birds (RSPB), Scottish Wildlife Trust (SWT), National Trust for Scotland (NTS), Woodland Trust (WT) and many others. (see *Useful Contacts* for addresses and functions of these bodies in relation to public access)

Despite their separate organisations and different remits, a characteristic of the conservation agencies in Scotland has been their capacity to communicate readily between organisations and to work together in a variety of partnership projects, such as Farming, Forestry and Wildlife Groups (FFWAG), Species Action Programmes, and Access Forums. This has been considerably assisted by the work of 'umbrella' groups such as Scottish Wildlife and Countryside Link and the Scottish Countryside Activities Council.

Perhaps the other distinguishing feature of conservation philosophy and action in Scotland is the recognition of the place which the natural

Farming for Fish

One of the most recent developments based on a valuable Scottish natural resource is that of fish farming, mainly of salmon, although shellfish farming is now becoming increasingly important. Because much of this takes place in the famed and beautiful sea lochs of the west coast and islands, and in many cases is the only obvious commercial development in these landscapes, concern has been expressed about its impact on the environment. The output of salmon alone was estimated at over 100,000 tonnes in 1997, from a figure of only one tenth of this some 10 years previously. The prediction is that output will rise to 132,000 tonnes by the year 2000. At present over 1,000 full-time jobs are involved in this sector. However the Scottish Environment Protection Agency has issued severe warnings to the industry that in respect of their use of chemicals - for example to control sea lice - some farms were putting more nutrients and trace elements into the marine environment than towns of tens of thousands of people, and that it must clean up its act.

Fish farming includes freshwater sites for both native brown trout and the introduced rainbow trout – there are fears that the introduction of the latter to boost stocks may endanger native fish populations.

Shellfish farming appears to pose far fewer environmental problems than fin fish farming - shellfish for human consumption are absolutely dependent on the highest water quality standards, and the impact on the landscape is much less obtrusive than salmon cages. There are now plans to grow freshwater crayfish in the warm water from distilleries, for example from the Talisker Distillery at Carbost on Skye, where not only could they grow all the year round in this warmed water from the distillery effluent, but they can also be fed on the spent barley grains from the distilling process. Shellfish include mussels, and it is appropriate here to refer to the freshwater or mussel, famed for the production of Scottish pearls, (for which it said that the Romans came north) and whose larvae thrive in the gills of sea trout. The species is so endangered from over collection that pearl fishing has now been banned.

characteristics of the country – its landscape and wildlife – has in contributing to a sense of national identity and pride, as part of the total heritage, embracing history, culture, and man's struggle, often under hostile environmental conditions, to live a full life. There is therefore an awareness that the land is a resource for survival, which goes beyond viewing the countryside simply as an add-on amenity for the pleasure and recreation of urban dwellers. One might go as far as to say that a high proportion of Scots, urbanites or not, have an unconscious affinity with their countryside, which is at least in part derived from its accessibility, and therefore have a sense of 'ownership' comparable to, and inseparable from, to their own roots and image of themselves.

CHAPTER 9

Protected Areas

THERE IS AN ALMOST BEWILDERING variety of different sorts of areas protected for their wildlife, scenic or recreational values in Scotland, from those officially designated by law to many managed as nature reserves by several voluntary organisations. As a result it is estimated that as much as 20 per cent of Scotland's countryside is specially conserved in one way or another. A number of these designations overlap: for example, National Scenic Areas may embrace National Nature Reserves, Sites of Special Scientific Interest and a number of voluntary body reserves.

Among the oldest forms of statutory conservation areas are **National Nature Reserves,** established under the 1949 National Parks and Access to the Countryside Act (although Scotland as yet has no national parks – see *By Yon Bonnie Banks*) these are now administered by the official government nature conservation agency, Scottish Natural Heritage. Such areas can range from a small peat bog of interest largely to scientific specialists, to very large multi-purpose areas such as the Cairngorms, or remote island groups such as St.Kilda. They include the first National Nature Reserve to be established in Britain at Beinn Eighe in Wester Ross, and the most southerly in Scotland at Caerlaverock on the Inner Solway, where controlled wildfowling is a key element of management. All of the reserves named above and many others are described briefly under *Places to Enjoy Wildlife.*

It may come as a surprise to hear that, despite their title, less than one-third of the total area of National Nature Reserves are actually owned by the state in Scotland – by far the largest area is under some form of management agreement negotiated with their individual private owners, of which there may be several on any one reserve. This means that while nature conservation is obviously an important element in the aims of management, many other interests may have to be accommodated, including agriculture, forestry, gamesport and recreation. In a number of cases, it may involve restrictions on public access, although many owners have agreed to the Access Concordat described below. Dependent on their ownership and vulnerability – and not all nature reserves are suitable for public use – access can range from areas which can only be visited with special permission (although this is very much the exception) to traditional recreational areas where there is completely open access.

Suitable reserves may have highly developed public access and viewing facilities such as Vane Farm just outside the Loch Leven National

Nature Reserve, or the educational visitor centre at Sands of Forvie. In a number of cases, these facilities are provided by voluntary organisations such as the RSPB or the National Trust for Scotland. The largest and most popular areas will have field staff or rangers (*see A Ranger at Your Service*) at least during the main visiting season, while others will have information boards and self-guiding trails. Increasingly, visitor centres are being made accessible to wheelchairs and in a few cases, the shorter trails may be suitable for the disabled.

Apart from their natural history interest, which can include everything from ferns and mosses to spectacular seabird colonies, National Nature Reserves as well as a number of other voluntary body reserves, are often excellent places to visit to see conservation in action. The biological records on such areas are amongst the most complete in the country and a number of areas have been the test-beds over a several years for experiments in nature conservation management. These include for example, the island of Rum, where the red deer herd has been culled at different levels to assess the effects on its grazing habitat and where the white-tailed sea eagle was re-introduced (*see The White-tailed Sea Eagle*) Elsewhere, as on Ben Lawers, protection of the arctic-alpine flora has involved fencing off of important areas, while on the Cairnsmore of Fleet in Galloway, traditional sheep grazing and burning has been employed to manage the vegetation. Some of these early experiments, such as the wildfowling scheme at Caerlaverock, have provided models for application outside Scotland.

It has to be said that, as elsewhere in UK, the designation of SSSI and their implications for development and land use, following the passage of the 1981 Wildlife and Countryside Act, have not been without consider-

---

PROTECTED LAND IN SCOTLAND

Scotland has no national parks but it does have:

*   70 National Nature Reserves - Beinn Eighe being the first declared in Britain while the Cairngorms is the largest nature reserve in Europe - most of which are accessible to visitors.

*   40 National Scenic Areas covering over 1 million hectares or nearly 13% of the land surface of Scotland.

*   Over 1,400 Sites of Special Scientific Interest including geological, botanical and zoological areas covering over 11% of Scotland.

*   The official government agency responsible for managing or advising on these areas and on access and recreation in the countryside is Scottish Natural Heritage which has area offices throughout the country.

*   In addition there are 6 Forest Parks with visitor facilities administered by the Forestry Commission covering approximately 200,000 hectares.

able controversy. Given that the majority are on private land, owners are understandably concerned that such designations might reduce land values and inhibit changes in management and use, which must now be notified to Scottish Natural Heritage. It happens that the largest areas under this form of protection are in the remoter districts of the Highlands and Islands, where both local authorities and farmers or crofters are concerned that, as they see it, they are being required to carry the heaviest burden of conservation, and where employment is often most difficult. There are criticisms that the system of designation is bureaucratic and undemocratic, with inadequate mechanisms for appeal. Notwithstanding the provision for compensatory payments, owners and others still feel financially disadvantaged. The fact remains that any system of conservation outside state-owned land is bound to impose constraints on land use and management if it is to be effective, and there is no doubt that this form of protection makes its own contribution to the generation of income from tourism and recreation which is attracted to these areas because of their natural beauty, albeit that not everyone in local communities may benefit directly from these activities.

Local Nature Reserves, which usually have bye-laws to prevent or restrict certain activities, are established and administered by the Local Authorities under the same Act as that for National Nature Reserves. They are sites of high conservation interest which in addition, normally have value for education and informal recreation by the public. A good example is Aberlady Bay on the East Lothian coast – the first such reserve established in Britain.

All National Nature Reserves and most Local Nature Reserves are designated as **Sites of Special Scientific Interest** under the Wildlife and Countryside Act of 1981. This is now the most important protective designation for nature conservation in the country. Covering over 11 per cent of Scotland (see *Sites Designated for Conservation*) such sites include a very wide range of situations. They may be tiny or huge, and include every natural habitat from coast to mountain top. Approximately one third of all such sites are exclusively geological in their interest. SSSI are required to be notified to the relevant local authorities and all owners and occupiers involved, as well as other statutory agencies in order that any proposed development or any potentially damaging operation is notified to Scottish Natural Heritage. (In the year 1996-97 SNH received 458 such notices of intent to carry out a potentially damaging activity on SSSI). Despite their importance for geological and wildlife conservation, the public is rarely aware of their existence, since they are not normally advertised and they are in most cases privately owned. Nevertheless, it is this provision which underpins the protection of the most important areas for conservation in the country, and which is a basic requirement of any such areas for their listing and proposed designation as international sites for their habitats or species under the various conventions and directives indicated below.

A substantial number of sites in Scotland are known to be of such conservation interest that they are listed in one or other of the various international conventions of European Directives which are becoming increasingly important in framing the necessary standards and regulations for wildlife in individual signatory countries. Some of the most important are indicated in the table below, from Special Protection Areas to Ramsar sites. (This might also have included the solitary World Heritage Site of St. Kilda.) Two of these categories – Special Protection Areas and Special Areas for Conservation – are grouped under **Natura 2000** which is the name given to the programme of work linked to the European Directives on Habitats and on Wild Birds, to secure the protection of a network of natural heritage sites of European importance. In Scotland at the present time, such sites are being given the highest priority for designation and protection.

National Scenic Areas (NSAs) are in a different category from those above in that they are not so much concerned with wildlife values as with landscape quality. They are the largest of protected areas, the most extensive covering the wild and rugged mountain area of the north-west Highlands, and are Scotland's only formally defined landscapes. Their purpose is to ensure that any development within them is properly controlled to reflect the highest standards in such valuable environments. For example, high level vehicle tracks are generally prohibited in such areas, and Scottish Natural Heritage must be consulted on any proposal for residential building schemes over a certain size. Local authorities are required to include appropriate policies for the conservation of NSAs within their development plans. The areas have been selected mainly on that combination of features which are most frequently regarded as beautiful. On the whole, this means that richly diverse landscapes which combine prominent landforms, coastline, sea and freshwater lochs, rivers, woodlands and moorlands with some admixture of cultivated land are generally the most prized. Because of the distinctive scenery in the south of Scotland, somewhat different criteria have been used to characterise the more subtle managed landscapes to be found there. Several of the NSAs have been identified as likely contenders for the first National Parks in Scotland (*see below*)

Scotland is almost unique among western nations in not having a system of National Parks which are therefore missing from the list of designations below. Unlike England and Wales, the provision for National Parks in the original 1949 Act did not extend to Scotland, for a variety of reasons. Among these was the notion that so much of Scotland was of outstanding scenic quality, that not only would it be difficult to define such areas, but that with its relatively low population, such protection was not required or demanded. There is no doubt that powerful landholding and

'By Yon Bonnie Banks......'

Loch Lomond, situated within one hour's drive for over 70 per cent of Scotland's population, is the first proposed National Park in Scotland to be established by the new Scottish Parliament. The loch is the largest freshwater body in Britain in terms of surface area, at 71 square kilometres and reaches a depth of 180m. It is unique in having two quite different ecological water systems within the one loch - a deep, cold, typically highland loch at the northern end, and at the other, a shallow, rich lowland type, separated by the most important geological division in Scotland, the Highland Boundary Fault. It is hardly surprising that with this range of conditions, the loch and its immediate vicinity supports approximately a quarter of the total number of different flowering plant species in Britain. It has no less than 19 different species of fish, including the very rare powan (a member of the herring family). Loch Lomond and its environs amply demonstrate the variety of different overlapping conservation designations which can be present in one area, lying within a National Scenic Area, part of which is both a Regional Park and Forest Park while a National Nature Reserve has been established over the islands in the middle of the loch and over the delta of the River Endrick at its southernmost end, not to mention many Sites of Special Scientific Interest!

development interests were also against National Parks in Scotland. Other opponents, strangely enough, included some countryside lovers, notably ramblers and climbers who feared not only 'park' type development, but paradoxically, a restriction on their claimed 'right to roam (*see below*)

Perhaps above all, there was the fear, reinforced by the experience of National Parks in England and Wales, that such labelling would simply attract more and more people, creating problems of congestion and damage in popular areas. Most of these arguments are at best unproven, but coupled with the reluctance of successive governments to face the additional costs of protecting and managing such parks, Scotland has had to wait until 1997 for the first tentative steps to be taken to establish such a system with the announcement of the proposal to create Scotland's first National Park over the area of Loch Lomond and the Trossachs. This has come about not only because of substantial public demand, but also of increasing recognition of the problems of fragmented administration, often with several competing Local Authorities in the same area, inadequate strategic conservation policies, and the lack of integrated management for our finest landscapes and wildlife conservation areas, such as for example, the Cairngorms.

Despite the rather bewildering plethora of different national and international designations, with their equally mystifying acronyms, what such labels highlight is that Scotland still provides an environment which has very fine landscapes and is rich in wild plants and animals, a number of these occurring in habitats which on a European or even global scale are unusual and sometimes unique.

SITES DESIGNATED FOR CONSERVATION *

Designation	Number of Sites	Area (hectares)

National Nature Reserves 70 113,238

Statutory reserves of national importance established under Section 19 of the National Parks and Access to the Countryside Act 1949 or Section 35 of the Wildlife and Countryside Act 1981 (All NNR are also notified as Sites of Special Scientific Interest -see below).

Sites of Special Scientific Interest 1,433 914,029

Sites of Special Scientific Interest are exemplary places in Scotland for nature conservation. They are also special for their plants or animals or habitat, their rocks or landforms or a combination of these. Designation is a legal process under the 1981 Wildlife and Countryside Act.

Ramsar Sites 35 69,192

Ramsar Sites are sites designated under the Conventions on Wetlands of International Importance especially as Waterfowl Habitat. The Convention was adopted in Ramsar, Iran in 1971 and ratified by the UK Government in 1976.

Special Protection Areas 67 115,454

Special Protection Areas (SPAs) are areas classified under the European Community Council Directive on the Conservation of Wild Birds.

Proposed Special Areas of Conservation are areas to be designated under the European Community Council Directive on the Conservation of the Natural Habitats and Wild Fauna and Flora (The Habitats Directive). There are 108 such proposed areas in Scotland

Biosphere Reserves 9 28,768

Biosphere Reserves are nominated by national governments for inclusion in the 'UNESCO Man and the Biosphere Programme' set up in 1971 to co-ordinate understanding of man's influence on the natural environment. They have no statutory basis, but all such areas in Scotland are also National Nature Reserves.

Biogenetic Reserves 2 2,368

A network of reserves to conserve representative examples of European flora and fauna and natural areas, primarily for biological research, under the Council of Europe's Berne Convention (ratified by the UK Government in 1983).

National Scenic Areas 40 1,001,800

National Scenic Areas (NSAs) are nationally important areas of outstanding natural beauty and represent some of the best examples of Scotland's grandest landscapes, particularly lochs and mountains. Such areas are protected through the

There is now some doubt, with much evidence of hybridisation with feral cats, as to whether the Scottish wildcat, noted for its ferocity, is truly a distinct species.

Pine martens have increased considerably in numbers in
recent years, especially in the Great Glen, and like the fox,
are now taking advantage of human food scraps.

With a strong preference for coniferous woods, the red squirrel has declined in numbers outside the Highlands of Scotland, and is threatened by competition from the introduced grey squirrel.

The dog-like face and smaller size of the common seal shown here distinguishes it from the more numerous grey seal, but both are frequently seen around the Scottish coasts, especially in the north and west.

At one time thought to be in serious decline in the lowlands, the playful otter is now found in most districts in Scotland. Otters are known to be considerable wanderers.

The Moray Firth is home to the only resident bottlenose dolphins in north-west Europe. They are far more numerous than anyone had previously realised, regularly seen from the shore.

The hen harrier, a target for gamekeepers, is easily recognised by its erratic low swooping flight over any moorland where voles or young grouse can be hunted.

Within Europe, Scotland is a stronghold for golden eagles, which, although they feed on carrion as much as on live prey, are still persecuted on sporting estates.

advice to Local Authorities contained in Circular 9/87 and Scottish Natural Heritage is required to be consulted on specific categories of development proposed within NSAs.

Local Nature Reserves   24   7,984

Local Nature reserves are sites which, in a local context, are of high conservation interest or of a high value for education and informal enjoyment of nature by the public. They are established by Local Authorities under Section 21 of the National Parks and Access to the Act 1949.

Regional Parks   4   86,125

Regional Parks (designated under section 48A of the Countryside (Scotland) Act 1967) are extensive areas of the countryside which provide for the co-ordinated management of facilities for informal recreation alongside and in close collaboration with the management of other land uses in the area.

Country Parks   36   6,426

Country Parks are relatively small areas of countryside near to the towns which are managed for public enjoyment. A variety of facilities are provided for informal recreation such as picnicking, walking and water activities, mainly located in the Central Belt, with the majority owned or managed by local authorities.

Source: Facts and Figures 1996-97 – Scottish Natural Heritage 1997

CHAPTER 10

Protection of Species

VISITORS FROM OUTSIDE BRITAIN may be surprised that with its wealth of game, from wild deer and salmon to grouse and wildfowl, there is no single official agency (comparable for example to the US Fish and Wildlife Service) charged with the responsibility of administering the game laws and regulating bag limits or the number of sport fish which may be caught (notwithstanding the particular responsibilities of specific bodies such as the Red Deer Commission and District Fishery Boards). Despite this, there is little evidence that game is overshot or that in general wildlife suffers directly from hunting pressures (whatever one's personal views about the ethics of this activity). Partly this is due to the fact that, coarse fishing apart, hunting and shooting has been confined by tradition and socio-economic factors (including ownership of game by the owner of the land involved) to limited groups of people. The one important exception to this is wildfowling which in Scotland is available to anyone on the foreshore between high and low watermark, and there is no doubt that in unregulated areas, wildfowl can suffer from disturbance from irresponsible shooters.

Game laws in Scotland are often of considerable antiquity, derived originally from the very restrictive laws imposed to protect game species for royalty and the nobility, especially in the extensive royal forests, often to the detriment of farmers and tenants unable to take action against marauding animals on their crops. One very clear exception to this from mediaeval times onwards was the wolf (see *Bring Back the Wolf?*), against which every man's hand was turned, partly because of its depredations on domestic stock, but also because of its reputation as a grave robber or even a molester of humans.

Many sporting estates were created in the 19th century when large areas of the Highlands of Scotland came to be regarded as a hunting sportsman's playground, at least for those with the means and social status to be able to indulge in deer stalking, grouse shooting and salmon fishing. All of these symbolise the distinction of field sports in Scotland, attracting hunters from many countries – and it has to be said, generating much-needed jobs and economic activity in rural areas where this is often lacking. Moreover, this form of land use can encourage the creation of habitats, including woodland and scrub cover, which is incidentally beneficial to other forms of wildlife, including small birds and mammals. However, there is good evidence that Scotland is only now recovering from the massive slaughter of wildlife species thought to compete with favoured game species – a toll which included wildcat, pine marten, polecat and all birds of prey. The game books up to the beginning of the First World War testify to the scale of this deliberate persecution. Unfortunately, despite a considerable improvement in this situation since these times, the recent recovery of birds of prey, possibly combined with economic pressures on sporting estates, have encouraged some unscrupulous owners and estate staff to defy the law on the protection of birds and to resume the practices of the past, now often using modern poisons to target golden eagle, buzzards, hen harriers and others. There is good evidence to show for example, that in the last 6 years over a hundred golden eagles have been killed illegally by gamekeepers and egg collectors, with the result that while other raptors appear to be increasing, the golden eagle population has remanied static at about 420 pairs. Nor are these the only birds to be targeted: research at Aberdeen University has concluded that over the last 5 years 1,667 goosanders, almost 1,000 red breasted mergansers and around 500 cormorants have been killed under licence from the Scottish Office Agriculture Fisheries and Environment Department, in response to demands from sport fishing interests. There is concern that these levels of killing could endanger Scottish populations of these birds.

The Law

Most plants, animals and birds, except those classified as weeds or pests, are generally protected by law in Scotland, which is little different in this respect from the law elsewhere in UK. In addition there are particular provisions relating to specially protected species and additional penalties for infringements regarding these. For example, it is not only an offence to kill, injure or capture any specially protected animal (eg bats) but also to disturb them at their place of shelter. Likewise, while all birds are generally protected from disturbance or collection of their eggs, there are special penalties for collection of eggs or disturbance of particular species, such as the great northern diver. Many people may not realise that its is an offence to uproot any wild plant. Although it is not a legal offence to pick (as opposed to uproot) wild flowers unless specially protected, people are advised not to do so. A licence is necessary to photograph at the nest any one of approximately 100 species of protected birds.

Game species such as grouse, ducks and geese may be protected for part of the year but can hunted at other times. It is an offence to release into the wild any wild animal or bird unless it is ordinarily resident in this country. The most important regulations are contained in the Wildlife and Countryside Act of 1981, but in addition there are several acts of Parliament relating to particular species, such as the Protection of Badgers Act 1992 and all wild mammals now have protection against cruel acts under the Wild Mammals Protection Act 1996.

Vulnerable Islands

Scotland has about 800 islands ranging from isolated archipelagos, such as the Shiants, to substantial land areas such as Skye. Island systems world-wide have for some time been recognised as important ecological 'outliers' with their own biogeographical and genetic reservoirs. In Scotland, St.Kilda is distinguished by having its own sub-species in the St.Kilda wren and the St. Kilda mouse. Conversely, a number of these islands are lacking in species which are a normal and common component of mainland wildlife, whether plant or animal. This is especially important in relation to aggressive or predatory species where potential prey have expanded in their absence. For example, neither stoat nor weasel are present in the Outer (Hebrides) Isles, nor on most of the Small Isles, where the badger and adder are also absent. Evidence from other island systems, such as the Galapagos, indicate that island flora and fauna develop in particular ecological and genetic ways in the presence or absence of competing or predatory species.

It is not surprising therefore that introductions, whether by accident or design, can cause severe disruption to island ecosystems. This has been dramatically demonstrated by the introduction of two highly predatory species, the American mink and the common hedgehog, both of which have wreaked devastation on the ground-nesting populations of birds for which the Outer Hebrides in particular are renowned. Mink have either escaped from mink farms as a result of inadequate containment or as a result of abandonment of unprofitable furfarms. Hedgehogs have almost certainly been deliberately – if innocently – released. With its capacity to swim, mink have been implicated in the massive reduction in water voles in recent years, and the decimation of gulleries and terneries, as well as reducing waterfowl populations, not to mention domestic poultry. There had also been fears that it might oust the native otter from its territories. However, the latest research has revealed a surprising situation. It now appears that in otter areas, mink are disappearing dramatically, and since they do not appear to be going anywhere else, the strong inference is that they are being killed by otters – not for food, but simply because they are competing for territory. The otter at ten times the weight of a mink would of course be quite capable of doing this.

Perhaps more surprisingly, the beloved hedgehog is now known to be the other villain of the piece in ravaging the nests of breeding waders in the Uists, including dunlin, redshank and snipe. Over a relatively short period, the numbers of these birds successfully rearing young there has fallen dramatically. From an initial release of a few animals in 1974, there is now an estimated 5-10,000 hedgehogs spreading inexorably northwards into internationally important areas for such birds. There is reliable evidence to show that hedgehogs have reduced breeding success from a normal average of 75 per cent to just 12 per cent. The most recent proposals are for a programme of trapping and release of hedgehogs from the islands back on to the mainland, while there is no possibility that they can now be eradicated. Although this is an extreme example, it does emphasise the vulnerability of island ecosystems and underscores the necessity for greater caution in contemplating even re-introductions anywhere, and the justification for the law on all introductions into the wild.

Coping with Change

Many species have declined rapidly in recent years, and in 1995 Scottish Natural Heritage launched its Species Action Programme which aims to maintain or restore viable populations of our threatened species across their traditional range (A number of the species are featured in the special topic boxes or elsewhere in this publication) This co-ordinated approach to the conservation of threatened species may involve specific habitat protection, ecological research, land use control in particular areas, and even cultivating species and translocating into the wild where necessary. (An outstanding example is the successful re-introduction of the white-tailed sea eagle – see *The White-tailed Sea Eagle*)

While various human activities can be a direct hazard to wild species, (such as the introductions referred to under Vulnerable Islands), by far the most serious threat is from changes in land use and management leading to loss of their natural habitat. Most species of plants and animals in this country are quite specific in their environmental requirements, but the tendency of modern development, whether of agriculture, forestry or industry, is to reduce the natural diversity of the environment and therefore the range of ecological niches for a wide range of our native flora and fauna. In the arable lowlands, these impacts are little different from elsewhere in Britain, emanating largely from the intensification of modern agriculture with its high technology and chemicals. Thus farm ponds and mosses have been drained and reclaimed, with the loss of wetland plants and amphibians in particular, hedgerows removed, and little land is now left fallow – away from set-aside areas – to provide for the needs of wintering birds. Old farm buildings which at one time attracted barn owls have now been replaced by modern 'hangars', while the so-called weed species of the cornfields are a thing of the past due to effective herbicides. Many insects which provide food for birds and mammals have disappeared under the onslaught of modern insecticides.

Elsewhere, in the uplands, the biggest single obvious change has resulted from large-scale plantation forestry, mainly using exotic fast-growing conifers, with quite drastic prior land preparation including ploughing, drainage, and fertiliser application. While the young plantations can be a rich hunting ground initially for birds of prey and the commoner woodland birds and mammals, the rapid closure of the canopy excludes all but the most shade-tolerant of plants and animals. Because of its extent of relatively poor hill ground, Scotland has seen a much larger area of its uplands converted in this way, with consequent loss of other habitats, including a substantial reduction in the open heather moorland which characterised many areas in the past. In addition, there is now a general consensus, that at least from the point of view of nature conservation, many areas have been overgrazed by sheep and latterly by red deer, so that the regeneration of natural scrub and native woodland has been severely inhibited. Around the coast and in the sea lochs, the rapid expansion of fish farms has raised a whole new set of unresolved ecological problems (see *Farming for Fish*).

The good news is that not only have many of these issues been recognised, but that action is being taken by a wide variety of bodies, not least farmers, foresters, and other rural managers, together with the official and voluntary countryside agencies to tackle these problems. There are now a whole series of special incentive schemes, including Environmentally Sensitive Areas, to encourage owners to make positive provision for the landscape and nature conservation interests, while the forestry agencies have in recent years demonstrated a considerable shift in attitude towards recognising and providing for the needs of other interests, including for example, the re-structuring of native pinewood areas by the Forestry Commission to restore parts of the Caledonian forest to something like its old grandeur, especially within forest parks. The particular value of the estuaries of Scotland as a natural resource of great commercial and conservation value has been recognised in the coming together of the many different interests involved in the special Focus on Firths programme, while the need to rationalise the public needs for open air recreation and access is being tackled through the Paths for All Initiative and the Access Concordat.

Out and About in Scotland

MOST ACCESS TO THE COUNTRYSIDE is informal, and there are few defined 'Rights of Way'. Although the law of trespass is little different from elsewhere in UK, it is not a criminal offence and therefore prosecution is not involved, but a so-called 'trespasser' may be asked to leave the land by the owner or occupier. However there is a tradition of free access to the hills, and there is usually no difficulty provided that reasonable care and courtesy to other users is observed (the Country Code below is a useful guide). Special care needs to be taken during the lambing season (particularly with dogs) in April/May. From August there is deer stalking and grouse shooting on many moors, and for their own safety, visitors are advised to check on access with the local tourist office or estate. Almost all of the places described under Places to Enjoy Wildlife are administered by agencies or private owners who welcome visitors and have provided public access facilities and information services.

The particular issue of access to the hills has received special attention in recent years with the increasing number of people using this resource for informal outdoor recreation. This is now a major use or our mountain country, alongside traditional activities of hill farming, forestry, field sports and deer management, all of which are important providers of employment. It is now accepted that there is a common interest between all these land uses and a joint responsibility on all who visit and manage the hills to conserve wildlife and landscapes and to have

The Country Code

* Guard against all risk fire
* Keep dogs under close control
* Fasten all gates
* Keep to the public paths across farmland
* Use gates and stiles to cross fences, hedges and walls
* Take your litter home
* Help to keep all water clean
* Protect wildlife, plants and trees
* Take special care on country roads
* Leave livestock, crops, and machinery alone
* Make no unnecessary noise
* Enjoy the countryside and respect its life and work

regard for the welfare of livestock. In order to establish better understanding between the various interests ant to promote toleration and co-operation, a Concordat has been agreed between the various representative bodies who have now agreed that the basis of access to the hills should be:

* *Freedom of access exercised with responsibility and subject to reasonable constraints for management and conservation purposes.*
* *Acceptance by visitors of the needs of land management, and understanding of how this sustains the livelihood, culture and community interests of those who live and work in the hills.*
* *Acceptance by land managers of the public's expectation of having access to the hills.*
* *Acknowledgement of a common interest in the natural beauty and special qualities of Scotland's hills, and of the need to work together for their protection and enhancement.*

The Hillphone, a recorded message answering service was piloted in 1996 to give information on where stalking was occurring in three areas of Scotland from August to October. These were the Mamore-Grey Corries on the British Alcan Estate, the estates lying between Loch Dochart and Glen Lochay, and the estates in the North Arran Hills. The information avoided disappointment for hillwalkers in finding areas of the hill closed, and reduced the chance of stalking operations being disturbed. Such was the success that the Hillphone again operated during 1997.

CHAPTER 12

When To See Wildlife

A GOOD RANGE OF Scotland's wildlife can be seen throughout the year, perhaps especially in mid-spring and autumn. Spring is outstanding for sheets of colourful woodland plants, while late autumn to spring is excellent for viewing wildfowl and waders. The spectacle of the autumn and winter run of spawning salmon can be enjoyed from mid-October onwards right into January, dependent on water flows. Winter too is the time when animals and their tracks in snow are most easily seen in the short days. The early summer months are the time to see some of the most spectacular displays of flowering plants, including the first arctic-alpines and orchids, while butterflies can be seen throughout the summer months. Early autumn can be especially fine in the Scottish Highlands with the last of the heather contrasting with the leaves of the birches turning, bright rowan berries appearing, and the russet bracken providing a backdrop on the hills; later fungi are plentiful in the woods.

A more detailed calendar is given here for the osprey as probably the most popular bird for viewing during the breeding season.

The Osprey Season

Late April to early May	Ospreys arrive, lay eggs, incubation starts
Late May to early June	Hatching
Mid-June to mid-July	Young very visible at the nest
Mid-July to mid-August	Fledging and first flights
Mid-August onwards	More time spent away from the nest and parents leave.

A WILDLIFE VIEWING CALENDAR

The best times to see Scotland's wildlife throughout the year

	JAN	FEB	MARCH	APRIL	MAY	JUNE	JULY	AUGUST	SEPT	OCT	NOV	DEC
PLANTS												
Upland				■	■	■	■					
Arctic – Alpines						■	■					
Bell Heather								■	■			
Ling								■	■			
Blaeberry						■	■	■				
Woodland				■	■	■	■	■				
Bluebell				■	■	■						
Wood Anenome			■	■	■							
Thrift					■	■	■					
Scots Primrose					■		■					
Orchids						■	■	■				
Thistle							■	■	■			
Harebell							■	■				
ANIMALS & FISH												
Red Deer				■						■	■	
Red Squirrel										■	■	
Grey Seal										■		
Ceteceans (Whales, Porpoises, Dolphins)							■	■				
Common Seal						■		■				
Salmon									■	■	■	■
Mountain Hare		■	■									■
Badger			■	■								
Fox											■	■
Otter								■				

	JAN	FEB	MARCH	APRIL	MAY	JUNE	JULY	AUGUST	SEPT	OCT	NOV	DEC
BIRDS												
Breeding Seabirds				■	■	■						
Skuas				■	■	■	■	■				
Migrants					■				■	■		
Wildfowl	■	■	■							■	■	■
Waders								■	■			
Terns						■	■	■	■			
Woodland				■	■	■						
Corncrake					■	■	■					
Birds of Prey												
INSECTS												
Scotch Argus							■	■				
Mountain Ringlet						■	■					
Dragonflies					■	■	■	■	■			

The periods given above are average for a number of groups and species, but allowance must be made for differences between north and south, with the southwest of Scotland and the Inner Isles having the mildest climates, and the earliest plants. A number of species can be seen outside these periods, but the times given are when you are most likely to be able to view, for example, breeding terns, or birds of prey hunting. The times of salmon runs vary between different rivers - some east coast rivers have early runs of spring fish, while west coast rivers may have few fish before June or even July. Some species can be seen at any time of year, such as wild goats and pine marten, but the latter, like the badger, is mainly nocturnal.

CHAPTER 13

Sustainable Wildlife Tourism

Sustainable wildlife tourism is concerned to ensure that the wildlife you enjoy is still there for you and others next year and the year after.

Wildlife watching should be fun: there are some regulations to protect our wildlife but they are difficult to police and it is usually the responsibility of wildlife watchers to make sure that their actions don't harm the very wildlife they have come to see. Wildlife watchers are on the increase and more and better equipment and services are available to help them.

Over and above the messages of the type contained in the Country Code and the Access Concordat, wildlife watchers should be more aware than most of the impact of their activities. Here are some issues to consider – remember one watcher or party may have little impact on wildlife, but multiply that several times and the cumulative impact may well have considerable effect.

Watching wildlife is a wonderful experience. Sometimes people get caught up in the excitement and spectacle and unintentionally harm wildlife. Most animals and birds react with alarm and freeze or flee when approached by humans, although some like seals may observe from a distance and some like great skuas will attack! These reactions are stressful and cause the animal or bird to expend energy. Although the energy cost of a single disturbance is usually small, repeated disturbance when food is scarce, during poor weather or in the breeding season, may result in high energy costs the animal or bird can ill afford.

Birds

The sensitivity of birds varies between species and season. Nesting birds should always be viewed from a distance. Flocks of birds can be flushed from roosts, open wetlands, beaches and cliffs by human presence; they might have moved away anyway, but acting with discretion is always wise. Some birds will soon tell you if you are too close – alarm calls, repeated chirping and 'broken wing' distraction displays are obvious signs, as are the divebombing activities of lapwings and skuas and the dramatic charge of the capercaillie.

Habitats need to be conserved just as much as species and natural processes will take their toll. Wildlife watchers can unwittingly accelerate the process: for example, walking over puffin burrows may collapse them. Scraps of food from picnic lunches will decay but in the hills they can attract and sustain crows and gulls which will then prey on other vulnerable species in the vicinity.

Think carefully what you might do if you come across a rare breeding bird or migrant. An influx of other watchers might cause disturbance or damage to the bird or its habitat and create problems for the landowner. You might like to tell an official organisation like the RSPB or keep it secret and undisturbed. Organisations including the SSPCA and the RSPB will advise on injured birds and it is often best left to trained staff to capture the bird to avoid further injury.

Mammals

Binoculars and telescopes usually provide the best opportunity for close up viewing, although hides or sitting quietly in a secluded spot waiting for a mammal to come to you can be successful. Watch out for clues that you might be disrupting normal behaviour. Head raised and ears and eyes pointed in your direction are a likely sign that you have been spotted. Be especially careful around females with young, since disturbing them causes unnecessary expenditure of energy when they can least afford it. Most Scottish mammals are unlikely to attack but apparently cute helpless seal pups can deliver a nasty bite, and during the rut in autumn red deer stags have been known to charge humans if suddenly disturbed. A general

---

A Ranger at Your Service

Many authorities and agencies employ countryside rangers (or their equivalent field officers on nature reserves) to assist the public to enjoy the countryside in Scotland. These include local government authorities, Scottish Natural Heritage, Forestry Commission, National Trust for Scotland, Scottish Wildlife Trust and some private estates. There are 90 ranger services in Scotland, almost equally divided between the public and private sectors, and all are supported by Scottish Natural Heritage. Sometimes these rangers are associated with a particular area, such as a regional park or nature reserve, while others cover whole districts. Whatever their situation, they will be familiar with the opportunities for outdoor recreation and the wildlife interest of their area. They will be skilled in interpreting this and other features of interest, including farming, forestry, and other rural land uses. For the visitor, they can be invaluable in providing an introduction to the area through talks at visitor centres or guided walks, often including special nature activities for families and children. Official Countryside Rangers can be recognised by their distinctive white and blue circular badges.

Do make a point of enquiring at the local tourist information centre about the ranger facilities in the district you are visiting.

awareness of natural cover, wind direction and a subdued colour of outer clothing can do much to aid your chances of seeing wildlife.

Marine Mammals

Watching marine mammals from the shore is unlikely to compromise the welfare of these, our largest wild animals. However venturing onto the sea in a commercial vessel or on a recreational trip can have some consequences for marine mammals. Make sure the operator taking you out understands the need for sensitive behaviour and follows any local codes when watching seals, dolphins, porpoises and whales. Boats, personal water craft and canoes can all harass marine mammals and responsible, cautious behaviour is essential, particularly if the animals have young or are in an enclosed situation such as an estuary or bay. It is not advised to touch live or dead animals as infections can be transmittted to humans.

Fish

At the end of the year Scotland has many locations where Atlantic salmon are arriving to spawn. In small burns and streams these exhausted fish are making spawning redds. If you are lucky enough to witness spawning salmon keep low and back away if the fish abandon the redds. Take care not to kick up silt or trample the stream banks.

Tread Lightly

Our attractive species often live in harsh environments on the hills or on the seashore. Low temperatures, strong winds and shallow soils all lead to short growing seasons and vegetation vulnerable to trampling. Research shows that the passage of 50 people over a raised bog can cause a 50% reduction in vegetation cover and it can take two years for growth to regain its original vigour. Take care of where and how you walk.

People can be subjected to the unwanted attentions of wildlife. Midges are perhaps the best known, but parts of Scotland, usually in bracken or heather, can harbour ticks with a fondness for biting passing humans. Don't panic, check yourself over and remove any unwanted passengers. Occasionally they can transmit viral infections to humans and animals but this is still very rare.

You can help wildlife even when you are not outdoors watching. If you are staying in the area ask your host in the hotel, guest house or bed and breakfast if they do anything to help the wildlife of the area. Good operators will have made conspicuous efforts ranging from providing advice on wildlife watching opportunities in the area to reducing their own impacts through energy efficiency and good waste and water management. There is a scheme run by the Scottish Tourist Board which accredits tourism providers who improve their environmental performance.

You can go further and consider being less tied to the car and walk or cycle more frequently. Some excellent wildlife watching opportunities can be had from the ferries plying to Scotland's islands. The rail journey through Sutherland and Caithness provides a view of the unique and internationally important peatlands of the Flow Country that is completely unavailable to the car traveller.

Many areas have crafts and produce made locally: buying these can help the culture and the environment of the area.

Duncan Bryden
Manager, Scottish Tourism and Environment Initiative

CHAPTER 14

Wildlife on Film

PHOTOGRAPHING WILDLIFE CAN OFTEN be a solitary occupation and the rare occasions that nature photographers get together present an ideal opportunity for exchanging ideas and comparing the differences of how each works. After years of attending such gatherings I have come to realise that many of the nature photographers living and working in Scotland rarely travel abroad in search of subjects because they have a rich and varied choice available to photograph right on their own doorstep.

The presence of this great variety is explained by the diverse and exciting range of habitats to be found in Scotland and one of the most enjoyable aspects of photographing wildlife lies in simply being out in these different environments, learning how potential subjects live and fit into particular ecosystems. Indeed, to succeed in photographing wildlife relies more on having an intimate knowledge of the subject than the photographic equipment used to record it. Most modern cameras, lenses and films are all capable of providing quality results once a basic knowledge of good working techniques has been mastered. From then, just finding a given subject can be a very real problem. My own approach is to visit selected areas time and again, sometimes in different seasons, until I can confidently predict where I may expect to see my chosen subject. Often, while visiting these areas, I will come across other subjects which I may either photograph there and then, or add to my ever lengthening list of things to cover.

Even covering the life-cycle of a single species can take months or years, either because certain aspects are very fleeting and easily missed, or because I underestimate it's potential and think of other facets of its behaviour which really need to be recorded to tell a more complete story. Sometimes, and just when I think that I have done everything I can, I will spend that extra hour, day or week in the field to try to improve on what I already have. I also think that it helps to become highly selective before committing anything to film. I have for example, learnt not to attempt to photograph even rare subjects that are in anything but the best of condition, most appropriate lighting, or without a suitable background. Though rare, I may be lucky and suddenly find myself confronted with a situation where all of these essential elements come together as if by magic. Such opportunities are of course more likely to occur by simply spending longer hours out in the field. Really elusive subjects do require

The re-introduction of the spectacular white-tailed sea eagle has returned to
Scotland its largest bird of prey, now breeding successfully on the west coast.

The evocative night call of the reclusive corncrake may now be heard more frequently in the islands as a result of recent special management schemes to arrest its decline.

Scotland is the host to almost the total population of the specially protected barnacle goose which winters on the Solway and on Islay, where numbers have risen steadily in recent years.

The colourful puffins feed exclusively on sand eels and may be at risk from the intensive commercial fishing of this species in northern waters.

A displaying capercaillie defending its pinewood territory
can be an intimidating sight, and will even attack humans!

Here in white winter plumage, the ptarmigan, a member of the grouse family, is one of the hardiest birds in the world, and in the central Highlands is found mainly above 800m.

Taking its Latin name from the Bass Rock in the Forth,
Scotland holds a significant proportion of the total world
breeding population in its island colonies of *Sula bassana*,
the gannet.

The great skua, the pirate of the moorlands of the northern and western isles, often displays his white-banded underwings from the top of any prominent knoll.

a certain degree of commitment just to get any results at all and the longer I spend working on them without a result, the more determined I become not waste all the time and effort, I have invested.

Inevitably perhaps, the much maligned Scottish climate can affect the length of time it may take to photograph a particular species and I always allow generous deadlines when working with those which live in the wetter West Highlands and Islands. But this is not to say that good photographs cannot be taken in bad weather. In fact, it may be much more appropriate to photograph say, mosses in a peat bog or ferns in a western oakwood on a rainy day than on a sunny one. Again, very cold weather actually creates good opportunities for photographing birds and mammals by altering their feeding behaviour and forcing them to concentrate in particular areas. The movement of red deer off the hills to the bottom of the glens, or the vast flocks of geese that migrate to our shores each winter are just two examples. The key to working successfully in bad weather lies in devising ways of keeping the camera dry and steady enough to permit the use of the longer shutter speeds needed to compensate for the lower lighting levels. Apart from the possibility of injecting more atmosphere into the photographs, an ability to cope with adverse weather suddenly means that there are many more days on which it becomes possible to take photographs each year. Climatic problems aside, the Scottish midge is much more of a hindrance, particularly when working with mammals which are easily disturbed by strong-smelling insect repellent.

Good wildlife photography cannot be rushed; to do so could compromise the welfare of the subject and this should always be regarded as more important than the desire to take a photograph. Rare species are all protected by law and nature photographers are required to be licensed just to go near to say, the nest of a barn owl or the drey of a red squirrel. Obviously the risks are fewer away from breeding sites and with a little imagination and detective work, it is often possible to identify opportunities where potential risks are minimal. Birds and mammals in particular, will often become tamer in situations where they have had little, or no bad experiences of people. Each evening, throughout the west highlands for example, the once persecuted pine marten commonly visits garden bird tables for handouts of peanuts or jam sandwiches. Again, one spring morning in a native pinewood, you may just be fortunate enough to find one of the few male capercaillies that has lost any fear of humans and continues to defend and stand its ground as you approach ever closer. Exciting as these encounters may be, I often think that we take the commoner species very much for granted. If starlings or mallard ducks were very rare then I am sure that people would travel great distances just to see them. By looking at them closely, and for long enough, then it is perhaps not difficult to marvel at the beauty of the intricate details of their plumage. The same could be said for much of our native wildlife, yet it is generally only the commoner subjects that allow us to approach close enough to appreciate it. I am therefore convinced that the actual species being photographed is less important than how it is being photographed. All of which could explain why, I for one, derive just as much satisfaction from photographing raindrops on grasses as I do golden eagles.

Laurie Campbell
Wildlife photographer

CHAPTER 15

Places To Enjoy Wildlife

THE SITES REFERRED TO in the following section represent merely a selection of the many places where the plants and animals, including birds, can be seen most readily in Scotland. The first section is mainly restricted to nature reserves and parks which are administered by official or voluntary bodies primarily for conservation purposes and where public use is mainly free (commercial suppliers are listed separately). Private estates are included where there is also involvement in management of a conservation agency. There are in addition a small number of known viewing points which are not within designated reserves or parks. For convenience, all of these have been grouped into identifiable tourist regions, so that they can be cross-referred to standard guide books such as the Scottish Tourist Board's *Touring Guide to Scotland* which also contains details of the other visitor attractions to be found in the vicinity, including for example, parks and gardens. The general descriptive introduction to each region focuses on the main natural features as the context for the specific locations described, but also refers to additional sites which may be visited. A number of localities which are specifically described later are emboldened in the regional description for reference.

For more detailed information on particular sites and reserves, the more comprehensive guides listed under *Further Reading* should be referred to.

GO SAFE!

While many visitors will be content to follow a well-marked nature trail and to use the viewing hides provided, others will wish to strike off up the hills on their own, or explore for wildlife round our coasts. Remember that the weather can change quite suddenly and dramatically at any time of the year, and that especially in winter, the daylight hours are short. The following tips to 'Go Safe!" should be considered as appropriate

* Wear boots with good grips

* Be prepared for wet or cold weather and take spare clothes as necessary.

* Check outward walking time and leave a generous margin for return before dark.

* If going any distance, let someone know your route and expected time of return.

* Get a weather forecast and be prepared to be flexible if the weather looks like turning against you.

* Watch for signs of tiredness and take energising food with you.

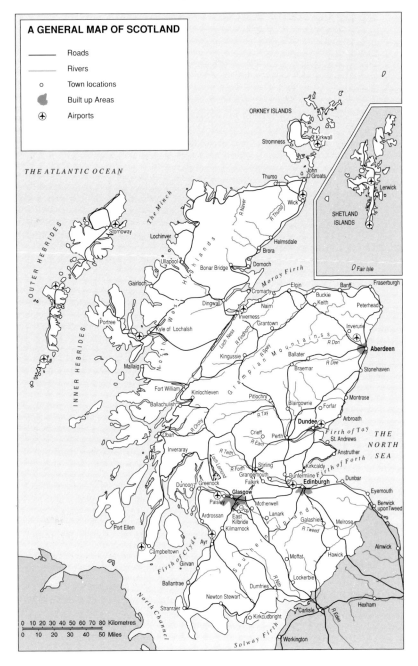

A GENERAL MAP OF SCOTLAND

——— Roads

············ Rivers

o Town locations

Built up Areas

⊕ Airports

THE ATLANTIC OCEAN

ORKNEY ISLANDS

Stromness

Kirkwall

Thurso

John O'Groats

Wick

Lerwick

SHETLAND ISLANDS

Fair Isle

THE MINCH

R Naver

R Thurso

OUTER HEBRIDES

Stornoway

Lochinver

Helmsdale

Brora

Ullapool

Dornoch

Bonar Bridge

Gairloch

Moray Firth

Cromarty

Elgin

Banff

Fraserburgh

Dingwall

Nairn

Buckie

Keith

Peterhead

North West Highlands

Inverness

Grantown

Inverurie

Portree

Kyle of Lochalsh

Loch Ness

R Findhorn

R Spey

R Don

Aberdeen

INNER HEBRIDES

Mallaig

Kingussie

Grampian Mountains

Ballater

R Dee

Stonehaven

Fort William

Kinlochleven

Braemar

Ballachulish

Pitlochry

Montrose

Oban

R Orchy

Blairgowrie

Forfar

Arbroath

R Tay

Crieff

Perth

Dundee

Firth of Tay

St. Andrews

THE NORTH SEA

Inveraray

R Teith

R Earn

Anstruther

Loch Lomond

R Forth

Stirling

Kirkcaldy

Firth of Forth

Dunoon

Greenock

Grangemouth

Dunfermline

Edinburgh

Dunbar

Paisley

Falkirk

Glasgow

Eyemouth

Motherwell

Berwick upon Tweed

Ardrossan

East Kilbride

Lanark

Galashiels

Melrose

R Clyde

Kilmarnock

R Tweed

Southern Uplands

Port Ellen

Ayr

Alnwick

Campbeltown

Girvan

Moffat

Hawick

Firth of Clyde

Ballantrae

Lockerbie

Newton Stewart

Dumfries

R Nith

Hexham

North Channel

Stranraer

Kirkcudbright

Carlisle

R Eden

0 10 20 30 40 50 60 70 80 Kilometres

0 10 20 30 40 50 Miles

Workington

Solway Firth

South of Scotland

1. **Dumfries & Galloway Tourist Board**
   Campbell House,
   Bankend Road
   Dumfries DG1 4TH
   Tel: 01387 250434 Fax: 01387 250462

2. **Scottish Borders Tourist Board**
   Tourist Information Centre
   Murray's Green,
   Jedburgh TD8 6BE
   Tel: 01835 863435/863688
   Fax: 01835 864099

3. **Ayrshire & Arran Tourist Board**
   Burns House, Burns Statue Square
   Ayr KA7 1UP
   Tel: 01292 262555
   Fax: 01292 269555

West Highlands & Islands, Loch Lomond, Stirling & Trossachs

4. **Argyll, the Isles, Loch Lomond, Stirling & Trossachs Tourist Board**
   Information Centre
   7 Alexandra Parade
   Duncon
   Argyll PA23 8AB
   Tel: 01369 701000 Fax: 01369 706085

5. **Kingdom of Fife Tourist Board**
   13-15 Maygate
   Dunfermline
   Fife KY12 7NE
   Tel: 01383 720999 Fax: 01383 625020

6. **Perthshire Tourist Board**
   Administrative Headquarters
   Lower City Mills, West Mill Street
   Perth PH1 5QP
   Tel: 01738 627958
   (Activity Line) 01738 444144
   Fax: 01738 630416

Perthshire, Angus & Dundee & The Kingdom of Fife

7. **Angus & City of Dundee Tourist Board**
   4 City Square
   Dundee DD1 3BA
   Tel: 01382 434664 Fax: 01382 434665

The Highlands & Skye

8. **The Highlands of Scotland Tourist Board**
   Information Centre, Grampian Road
   Aviemore, Inverness-shire PH22 1PP
   Tel: 0990 143070 Fax: 01479 811063

Grampian Highlands, Aberdeen & The North East Coast

9. **Aberdeen & Grampian Tourist Board**
   Migvie House, North Silver Street
   Aberdeen AB1 1RJ
   Tel: 01224 632727 Fax: 01224 848805

Outer Islands

10. **Western Isles Tourist Board**
    26 Cromwell Street,
    Stornoway
    Isle of Lewis HS1 2DD
    Tel: 01851 703088 Fax: 01851 705244

11. **Orkney Tourist Board**
    6 Broad Street
    Kirkwall
    Orkney KW15 1NX
    Tel: 01856 872856 Fax: 01856 875056

12. **Shetland Islands Tourism**
    Market Cross, Lerwick
    Shetland ZE1 0LU
    Tel: 01595 693434 Fax: 01595 695807

13. **Glasgow & The Clyde Valley**
    Greater Glasgow & Clyde Valley
    Tourist Board
    11 George Square
    Glasgow G2 1DY
    Tel: 0141 204 4400 Fax: 0141 221 3524

Edinburgh – City, Coast & Countryside

14. **Edinburgh & Lothians Tourist Board**
    Edinburgh & Scotland Information Centre
    3 Princes Street,
    Edinburgh EH2 2QP
    Tel: 0131 557 1700 Fax: 0131 557 5118

An overall description is provided of those main areas which are usually managed by a conservation body, whether official or voluntary, and which have both easy public access and visitor facilities, including appropriate information. This can range from quite sophisticated visitor centres with substantial exhibitions and audio-visual displays to sites which may only have a simple information board or leaflet. These main sites are indicated on the relevant regional map at the beginning of each section, and their location is briefly indicated. The district following the site name is the new local authority council area following the 1996 reorganisation of local government in Scotland.

Other sites within a reasonable distance of the main sites (but which may have minimal facilities and be less accessible) are provided with shorter indicative descriptions, the managing body, and a map reference for use with a suitable scale Ordnance Survey map, such as the standard 1:50.000 (2 cm to 1 km) Landranger series. The latter includes 2 alphabetical letters, indicating the south east corner of each 100 km square, followed immediately by blue west to east horizontal figures ('eastings') along the top or bottom of the map. The final 2 or 3 figures, dependent on whether a 4 or 6 figure reference is given, relate to the blue figures ('northings') on the sides of the map.

It is worth noting that whereas access to a visitor centre or short nature trails is usually without difficulty, further exploration of the interior of the larger reserves, especially mountain properties, may only be possible for those who are fit and properly equipped. Look out for notices detailing guided walks – for most visitors, these will usually provide an informative insight about the reserve in a relatively short time (*see topic box – A Ranger at Your Service*). Facilities for the disabled vary considerably, and even where the symbol indicates this, it may be restricted to access to a visitor centre.

The abbreviations for facilities at main wildlife sites are as follows:

L	leaflet available	I	information/visitor centre
R	ranger/warden	*	wheel chair access
WC	toilets		

Agencies responsible for managing natural areas

FC	Forestry Commission	SWT	Scottish Wildlife Trust
NTS	National Trust for Scotland	WT	Woodland Trust
RSPB	Royal Society for the Protection of Birds	JMT	John Muir Trust
SNH	Scottish Natural Heritage	HC	Highland Council

CHAPTER 16

The Highlands

NOW CONSIDERED TO BE one of the most extensive semi-natural areas in Europe, this is the region of Scotland dominated by high western mountain country with a varied geology based on some of the oldest rocks in the world. The Great Glen running south west from Inverness, and including the famed Loch Ness, is a major faultline dividing the north west Highlands from remainder of the region. At its southern end it is overlooked by the highest mountain in Britain, **Ben Nevis.** To the west of this the rainfall can range from 2,500mm to as low as 500mm in the east, while the Gulf Stream on the north western coast creates a surprisingly mild climate. It is this geology combined with plentiful rainfall which creates some dramatic gorges and waterfalls such as those at **Corrieshalloch Gorge** and **Falls of Glomach**

With some of the region having less than ten inhabitants per square kilometre, this is one of the least populated areas in Europe, partly the result of a hard environment, but also because of land clearance and emigration in the 18th and 19th centuries – there is a sense of emptiness emphasised by the many derelict farmhouses and crofts. However in Caithness and Easter Ross there is also rich, well cultivated farmland, and Inverness, the capital of the Highlands is a fast-growing commercial centre. Construction yards related to the oil industry along the shores of the Cromarty Firth and at Nigg are a relatively new phenomenon. Sheep rearing has declined in importance in recent years, but this is still the region of vast sporting estates dedicated to stalking and grouse shooting, although the quality of the latter has declined considerably this century. Fish farming is now a feature of most of the sea lochs on the west coast. Wildlife viewing and outdoor activities generally are well catered for in this region and are becoming an increasingly significant part of the tourism economy. This is perhaps the only region in Britain where visitors can have a real 'wilderness experience' by backpacking into the remote fastnesses of Letterewe and Fisherfield Forests, the **Rough Bounds of Knoydart,** or in the arctic wilderness of the **Cairngorms.**

Water is everywhere, from the great sea lochs such as Loch Torridon, Loch Broom, and Loch Eriboll, which penetrate far inland on the west and north coasts, to a myriad of tiny peaty lochans and fast highland burns. In the Flow Country of Caithness and Sutherland the vast undulating peat bogs stretching to the horizon are renowned for their breeding birds. The hard acid rocks generally produce poor thin soils and combined with high moisture levels, such peat is widespread. A very different sort of wetland is created around the upper Spey on the flooded **Insh Marshes** near Kingussie. Throughout the region there is evidence of extensive glaciation in the perched rocks and ice-scoured glens. However there are pockets of limestones such as at **Durness,** sometimes forming the characteristic pavements and with a rich flora.

In Wester Ross the Torridonian mountains drop sheer down to the sea and the scale of this landscape is awe-inspiring. The mountains of the west coast are typified by the wildness of **Ben More Coigach** north of Ullapool, the haunt of wildcat, golden eagle, red deer and many beautiful upland plants, while otters are not infrequently seen here as well as from the shore at **Balmacara**. The finest assemblages of high alpine arctic plants are to be found on Beinn Dearg and Seana Braigh south of Ullapool, while the isolated peaks of Suilven and Canisp are spectacular punctuation marks in the lunar landscape of this part of Sutherland. Large areas of open upland country have been afforested with plantations of fast-growing coniferous trees in recent years.

The region has some of the most remote and spectacular sea cliffs, such as those of the Clo Mhor on the north coast and **Handa Island**, with its great congregations of nesting seabirds. Elsewhere there are superb sweeps of sandy bays and dune grassland ('machair') such as at **Achmelvich and** Clachtoll. The area includes not only the largest of the inner Hebridean islands of Skye, but also the fascinating Small Isles of **Rum, Canna, Eigg** and Muck, all with their distinctive wildlife. On the east coast, sandy beaches give way to the offshore mudbanks of the estuaries in the Dornoch and Cromarty Firths, renowned as winter haunts for large populations of wildfowl and waders at **Udale** and **Nigg Bays**. The marine fauna, including whales, porpoises and dolphins are becoming increasingly well known as a wildlife spectacle, for example, at **North Kessock**.

While native deciduous woodlands are now mere remnants of their former extent, these still display fine assemblages of non-flowering plants such as mosses, ferns, and lichens characteristic of the moist conditions of the western highlands, and pinewoods such as **Beinn Eighe** and **Glen Affric** are renowned for their variety of typical wildlife amid spectacular mountain scenery. The distinctive drier eastern pinewoods are best represented by those in **Speyside**, such as **Rothiemurchus** and **Abernethy**. Not far away, at **Craigellachie** immediately above Aviemore, there is a fine example of a typical highland birchwood.

The large stately houses of the lowlands are absent here, but there are good examples of the 18th and 19th century planned townships of considerable charm such as Cromarty and Ullapool, and the preserved battlefield at Culloden is an important historic focal point. The military barracks at Fort George is the largest of its kind in the whole of Europe. Perhaps the most important prehistoric remnants are the brochs or fortified towers, which are found throughout this region, the best examples on the mainland being at Glen Beg, not far from Kyle of Lochalsh, and at Dornadilla, south of Loch Hope. In Caithness, between Latheron and Wick, the Grey Cairns of Camster are among the finest examples in the country of this particular form of burial chamber. The crofting system and highland agriculture, together with examples of related buildings and implements, is well displayed in the Highland Folk Museum at Kingussie, south of Aviemore.

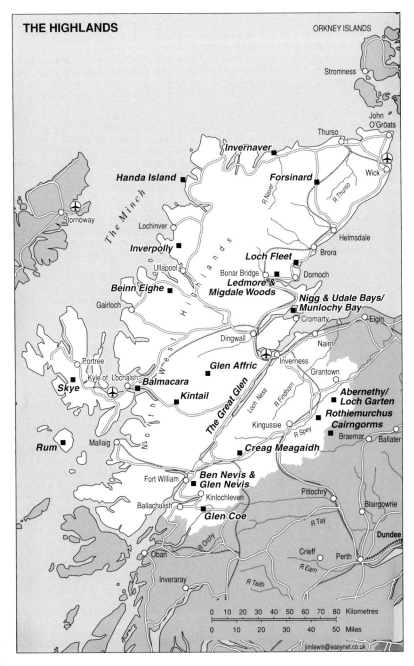

THE HIGHLANDS

ORKNEY ISLANDS

Stromness

John O'Groats

Thurso

Wick

Invernaver

Forsinard

Handa Island

R Naver

R Thurso

The Minch

Stornoway

Lochinver

Helmsdale

Inverpolly

Brora

Ullapool

Loch Fleet

Bonar Bridge

Dornoch

Beinn Eighe

Ledmore & Migdale Woods

Gairloch

Nigg & Udale Bays/ Munlochy Bay

Cromarty

Elgin

Dingwall

Nairn

Glen Affric

Inverness

Grantown

Portree

Kyle of Lochalsh

Balmacara

Abernethy/ Loch Garten

Skye

Kintail

The Great Glen

Rothiemurchus Cairngorms

Loch Ness

R Findhorn

Kingussie

R Spey

Braemar

Ballater

Rum

Mallaig

Creag Meagaidh

Ben Nevis & Glen Nevis

Fort William

Pitlochry

Blairgowrie

Ballachulish

Kinlochleven

Glen Coe

R Tay

Dundee

R Orchy

Oban

Crieff

Perth

R Earn

Inveraray

R Teith

North West Highlands

| 0 | 10 | 20 | 30 | 40 | 50 | 60 | 70 | 80 | Kilometres |

| 0 | 10 | 20 | 30 | 40 | 50 | Miles |

jimlewis@easynet.co.uk

HANDA ISLAND *Highland*

One of the largest seabird colonies in North West Europe, this reserve is remarkable for its Torridonian sandstone cliffs, in the north rising sheer from the Atlantic. Between April and July its huge colonies of breeding guillemots and razorbills are renowned, but there are many other common seabirds such as kittiwake, fulmar, puffin and shag, as well as great skua and arctic skua – the seabird colonies are seen in their most dramatic setting of the Great Stack of Handa. On the moorlands in the central part of the island, several species of orchids can be found, together with a variety of other bog plants, including the insectivorous pale butterwort, especially along the margins of the six lochans, which add variety to the grassland and heather of the interior. Other notable plants include Scots lovage and royal fern. There is a seasonal ranger service but the reserve is not especially suitable for small children or those with any walking difficulties – conditions underfoot require strong footwear. Visitors are restricted to a marked circular path and are asked to make a contribution to the cost of managing this spectacular offshore island.

3 miles north west of Scourie, and signposted from the A894. Access is by regular boat (except Sundays) across a mile of sea from Tarbet, which operates between mid-April and mid-September.

L R I *(SWT)*

Achmelvich *(HC NC 053243)* Beach, dunes, rich machair grassland and inshore seabirds.

Clo Mor *(NC 3272)* The highest cliffs on mainland Britain support huge numbers of seabirds, including one of the largest colonies of puffins in the country.

Eilean Hoan *(RSPB NC 4567)* Uninhabited island whose varied birdlife including great northern divers (spring) can be seen through binoculars from the coast road.

Durness *(NC 3871)* The walk from Balnakeil to the north-west of Durness village to Faraid Head is rewarding, with its limestone grasslands, high sand dunes and cliffs. Mountain avens, yellow saxifrage, alpine bistort, all grow here while this is a known site for Scots primrose where there is open sand between the grassland.

GLEN AFFRIC *Highland*

This is one of the most attractive locations in Scotland with its combination of hill, loch, river, islands and waterfalls and outstanding native pinewoods – there are many fine old trees between Loch Affric and Benevean, where the most natural areas of woodland occur. The extensive area supports most of the typical highland fauna, including red deer, wildcat, red squirrel, black grouse, capercaillie, and golden eagle – the larger birds of prey are frequently seen here, while the two birds especially associated with pinewoods, the Scottish crossbill and the crested tit, are often encountered along the path leading from the car park at Dog Fall.

Much of the fantastic scenery can be viewed from the road which winds along the glen, and is at its best in spring and autumn, with the vibrantly contrasting colours of pine, birch and rowan. The lochs and rivers support such bird species as the delicate grey wagtail and bobbing dipper, in addition to the common mallard and goosander on the lochans. Along the many winding forest tracks, typical western wet moorland plant species including abundant sphagnum mosses occur. There is an interesting forestry programme of restoring the old pinewoods from later plantings of exotic conifers, to create one of the largest continuous tracts of native Scots pine forest anywhere in the country.

A831 to Cannich and to head of Loch Benevean car park. Open throughout the year.
L R WC *(FC)*

Strathfarrar *(SNH NH 2737)* Fine native Scots pinewood, with some massive trees especially on the south side of the River Farrar, within a large National Nature Reserve.

Balmacaan Woods *(WT NH 500290)* Oak and birch woodland with pine plantations, with fine viewpoint at the north end overlooking Loch Ness.

Farigaig Forest *(FC NH 523237)* On the east shore of Loch Ness, a mixed deciduous and conifer woodland with a high viewpoint and forest centre display.

Falls of Glomach *(NTS NH 017258)* Possibly the most spectacular waterfall in Britain, with a drop of over 120m, situated in a remote mountain area for fit walkers only – see Balmacara.

THE GREAT GLEN *Highland*

The glen, about 100km long and following one of the great geological fault lines of Highland Scotland, divides the Highlands from coast to coast, high-lighted by a chain of long narrow lochs, dominated by Britain's highest mountain, **Ben Nevis** (see separate entry) at its southern end, while Loch Ness, the single largest volume of freshwater in Britain, lies at the opposite end. Representing a cataclysmic fracture in the earth's crust millions of years ago, it is still the most active earthquake zone in the UK. While the Fort William area is one of the wettest in the country, with an average of 2m annual rainfall, the vicinity of Inverness is one of the driest with only 0.65m. Most of the original ancient broadleaved woodland has long since gone, but parts of Farigaig Forest (see separate entry) at **Inverfarigaig** and **Horseshoe Crag** in Easterness Forest near Fort Augustus, are good examples of the surviving relicts. South of Fort William, above Loch Linnhe, **Ach an Todhair**, has birch, ash, rowan, hazel and supports no less than nine different orchid species. Above Alltsigh Youth Hostel near Invermoriston and in **Glen Moriston**, there are remnants of the old Caledonian pinewood, sheltering such typical plants as common and intermediate wintergreen – Glen Moriston is one of the last places where the capercaillie was recorded prior to its extinction in the mid-19th century. Red squirrel are relatively common and pine marten are regularly seen, especially in **Glen Urquart**. In the spectacular gorge at the **Falls of Foyers**, and similar ravines above **Loch Lochy**, roseroot and starry saxifrage thrive, among vivid green ferns.

Such areas, particularly at the west end of the Great Glen wherever shady glades are available, are home to the rare chequered skipper butterfly, as well as green hairstreak and Scotch argus. A quite different type of woodland occurs at **Urquart Bay** near Drumnadrochit, where the Rivers Coiltie and Endrick meet on Loch Ness-side. Here the waterlogged swamps have resulted in a tangled mass of alderwood, whose fringes favour orange tip butterflies. On the margins of **Loch Oich**, water lobelias, starwort and bogbean provide a colourful show in season, and the uncommon grasshopper warbler can be heard. By contrast, the bare shingle beach at **Aldourie** on Loch Ness has sea campion, herb robert and enchanter's nightshade. Arctic terns nest on these shingle banks, but where reedbeds occur, such as alongside the **Caledonian Canal** and along the **River Lochy**, sedge warblers, whinchat and reed bunting nest, while kingfishers have also been recorded here. Osprey are regularly seen in the glen. In these wet habitats, damselflies and dragonflies are common – 13 of Britain's 40 species have been recorded in the Great Glen, one of the commonest and largest being the aptly named golden-ringed dragonfly. Little grebe and Canada goose breed on **Loch nam Marag** near Inverlochy, and **Loch Duchfour** at the north end of Loch Ness, attracts goldeneye amongst a variety of other duck species. Perhaps the surprising ornithological element are the seabirds which use the glen as a flyway between the Atlantic and the North Sea, including guillemots and kittiwakes, probably blown off course, while fulmars have been seen flying over Inverness.

BEINN EIGHE *Highland*

The first National Nature Reserve in Britain, Beinn Eighe has an interesting visitor centre at Aultroy (open May to September) and a variety of attractive walking trails from the shores of the renowned Loch Maree to the mountain fastnesses above the woodland. The surrounding mountain and loch scenery is breathtaking, providing a great range of highland habitats and wildlife. Most of the mammals and upland and woodland birds of the Highlands are here, and there is a good prospect of seeing red deer, particularly in winter. This outstanding example of western pinewood, with high rainfall, is especially luxuriant with deep carpets of heather and other shrubs such as bilberry and cowberry under plentiful birch, rowan, as well as ancient pine, often dramatically seen against the steep cliffs of the lower mountain.

The ice-gouged corries in the ancient Torridonian sandstones are especially spectacular, set in the startling white of the quartzite scree below the main ridge of the mountain. The popular mountain trail displays the different climatic and vegetation zones up to 540m leading on to dwarf scrub of juniper, heather and bearberry, set in a mat of mosses and liverworts. On the higher slopes, mountain hare and ptarmigan are found. In richer pockets of soil, a variety of arctic-alpine plant species occur, such as mossy saxifrage, moss campion, mountain sorrel, Highland saxifrage and many others. Walking at these altitudes requires appropriate mountain equipment and fitness.

W of A896/A832 junction at Kinlochewe.

L R WC I *(SNH)*

Torridon *(NTS visitor centre at NG 905557)* Adjoining Beinn Eighe, a vast area of wilderness especially noted for its huge ice-carved amphitheatres of red sandstone rock at Coire Mhic Fhearchair.

Corrieshalloch Gorge *(NTS/SNH NH 204777)* Spectacular views down this fine wooded gorge and Falls of Measach can be obtained from the bridge and projecting viewpoint.

Rassal Ashwood *(SNH NG 8443)* Small National Nature Reserve on limestone noted for its uncommon lichens, particularly attractive with its flowering primroses in late spring.

Inverewe Gardens *(NTS NG 857821)* A good variety of woodland and shore birds.

BEN NEVIS/GLEN NEVIS *Highland*

The highest mountain in Britain at 1,344m, Ben Nevis shows a range of vegetation zones from virtually sea level to high montane on its huge dome of lava-crowned granite. Glen Nevis is one of the most spectacular glens in Scotland. Physical features which can be seen without hillwalking include one of Scotland's most impressive waterfalls in the Nevis Gorge at An Steall, over 1000m high, a dramatic waterslide careering for more than 350m down granite slabs from Coire Eoghainn in the upper glen, and gigantic potholes eroded in the river bed in its upper section. In the Nevis Gorge, a characteristic feature is the cushions of mosses and leafy liverworts which thrive under these damp shady conditions.

There are magnificent stands of relict pines in upper Glen Nevis, mixed with birches, rowan and holly, overlying blaeberry and heather, while alderwoods line the banks of the River Nevis. The lower slopes of Ben Nevis are dominated by grassland and heath vegetation, but above 450m true mountain species appear, including alpine lady's mantle and several species of saxifrage which can be seen beside the main path. Towards the summit plateau, tussocks of spiked wood-rush and parley fern also grow beside the path among the lava boulders. At the very highest levels, only a few mosses and lichens can survive. Here in summer, snow buntings are frequently seen as well as ravens and the occasional golden eagle.

Near Fort William

L R WC I *(HC)*

Loch Sunart *(SNH NM 8464 and 6558)* While the whole of Loch Sunart and its surroundings are noted for varied wildlife, the oak woodlands at Arriundle in Strontian Glen on the north side, and the ash and birchwoods of Glen Cripesdale on the opposite shore, are specially protected mainly for their ferns, mosses, and liverworts.

Rahoy Hills *(SWT NM 690535)* An exceptionally diverse range of habitats from exposed hill summits to freshwater lochans, mires, burns and mature oak woodlands, with guided walks from the Black Glen car park.

Glen Roy *(SNH NN 298854)* Well known to geologists for the striking demonstration of previous loch levels evident in the 'Parallel Roads' cut into the hillside of this geological reserve.

GLENCOE *Highland*

The visitor centre is a good starting point to explore this NTS property, made infamous by the massacre of the MacDonalds which took place in the 17th century. The area is described as a cauldron subsidence of volcanic lavas with large granite intrusions along the main fault line. Although the mountain flora is rich, it is usually to be found on the more inaccessible calcareous cliffs and away from the lower paths - much of this area is only for experienced climbers. However, after crossing the NTS footbridge over the River Coe near the Meeting of the Three Waters the steep path leading from the old road on the south side in the middle of the Pass of Glencoe takes the visitor over into the so-called 'lost' valley of Coire Gabhail where on the valley floor, mainly of shingle and grass in a magnificent alpine setting, can be found starry and yellow saxifrage, moss campion, alpine lady's mantle, roseroot and mountain sorrel. The dark brooding scenery is amongst the most atmospheric in Scotland, and the wild country is a haven for many mountain species including wildcat, eagle and ptarmigan.

For those who prefer the easy way to the heights, the chairlift at the ski-centre of White Corries at Kingshouse towards the eastern end of the pass takes visitors, after a short walk to the grassy summits, to some of the finest viewing points in the country, overlooking the famous isolated rocky peak of Buchaille Etive Mhor (Great Shepherd of Etive) to the south and the great expanse of Rannoch Moor (see description below) to the north-east.

The NTS visitor can be found on the A82 17 miles south of Fort William.

L R WC I *(NTS)*

Rannoch Moor *(SNH NN 4053)* The notorious bleak expanse of Rannoch Moor, extending into Perthshire, is one of the most extensive open boglands in Britain. It is a haven for wildlife, including the plants and birds of peatland and the many lochans here, in addition to herds of red deer and predatory birds.

Doire Donn *(SWT NN 050703)* A fine example of a north western oak and birchwood above Loch Linnhe, with unusual populations of insects, ferns and mosses .

Loch Ruthven *(RSPB NH 638281)* The most important site in Britain for the Slavonian grebe which may be seen from the reserve's hide, as well as osprey, on this peaceful highland loch.

RUM *Highland*

Rum is unique – a microcosm of a previous West Highland sporting estate, now an internationally famous nature reserve and a mecca for geologists: the complex of igneous rocks on the island provide an outstanding location for the study of basaltic volcanoes, while the unique bloodstone found here was prized by the earliest Mesolithic peoples for their tools. Apart from its wealth of natural history interest, Rum is the site of the earliest human settlements yet found in Scotland, dating from almost 9,000 years ago. Some 60,000 breeding pairs of Manx shearwaters – one of the largest colonies in the world – returning at dusk to their burrows high up on the hillsides and the roaring of the red deer stags in autumn, are just two of the experiences awaiting the visitor.

Now dedicated to active nature conservation as one of the few National Nature Reserves wholly owned by the Government conservation agency, the reserve was the launch pad – literally – for the re-introduction of the white-tailed sea eagle (see *White-tailed Sea Eagle*) but it has a wealth of other wildlife interest, including mountain plants and insects. It is also the location of a ground-breaking and very successful woodland restoration project using locally raised native trees – over 1 million planted to date. A centre for ecological studies over 40 years, Rum is especially well known for pioneering research on the ecology and management of its indigenous red deer herd. The unique fold of distinctive Rum ponies and the herd of pedigree Highland cattle – the latter used to maintain the quality of the flora of the grasslands – provide added interest. A variety of accommodation from castle hostel to bothy and camping is available, but the island is also accessible for day trips from Mallaig and Arisaig.(summer only)

Regular Caledonian MacBrayne ferry service (tel: 01687 462403) throughout the year from Mallaig. For accommodation contact Reserve Office on 01687 462026.

L R W C *(SNH)*

Eigg *(SWT NM 474875)* Part of the Small Isles (which includes Rum), this island has 3 nature reserves and a wide variety of plants and animals associated with its uplands, lochans, cliffs and bogs. In 1997 it was purchased by a trust, including the island community, for sustainable development and conservation.

Canna *(NTS NG 2705)* A farmed island bounded by distinctive stepped cliff coast, with a variety of breeding seabirds, several species of birds of prey, and abundant shoreline waders.

INVERPOLLY *Highland*

This rough wilderness surrounding the spectacular peaks of Stac Pollaidh, Cul Mor and Cul Beag, is a great natural sanctuary for no less than 104 bird species from coast to mountain top, including several birds of prey. A huge plateau of undulating ancient gneiss dominates the central area, covered mainly with wet heath of deergrass, heather, cotton grass and purple moor grass, with many bogs providing a habitat for the fragrant bog myrtle, while the intervening pools are showy with bogbean in season. Around the margins can be found the great sundew and common butterwort, contributing to the 360 plant species which this National Nature Reserve supports. Patches of scattered birch and hazel represent the previously extensive woodlands, but still contain a diversity of plants such as hay-scented buckler fern, lemon-scented fern, and melancholy thistle. It is also the home of wild cat, pine marten, and otter in its many lochs and islands, not to mention a 500-strong herd of red deer.

While the interior is only for the fit and well-equipped, the visitor centre and trail at Knockan Cliff are readily accessible, specialising in displaying the fascinating ancient geology of this part of the Highlands. The trail also allows the visitor to see a fine range of flowers on the limestone cliffs, including alpine lady's mantle, yellow saxifrage, mountain avens, alpine bistort, crowberry, and stone bramble. Visitors are warned that the popular walk up Stac Pollaidh ends in dangerous and crumbling rocks.

Signposted off A835 15 miles north of Ullapool.

L R WC I *(SNH)*

Benmore Coigach *(SWT NC 075065)* Large area of rugged mountain and wetland with typical upland flora and fauna – offshore is the separate reserve of Isle Ristol with machair and nesting seabirds (Access by boat from Old Dornie). The mountain area is dominated by heather and grass moorland, interspersed with birchwoods, where much of the botanical interest lies and in the crags and higher rocky areas, with brittle bladder fern, alpine scurvy grass, trailing azalea and dwarf juniper. The birdlife is diverse from ptarmigan and raven on the heights to ring ouzel and greenshank lower down.

Inchnadamph *(SNH NC2719)* Limestone plateau and a fine new trail to the Allt nan Uamh caves which have provided important relicts of human and now extinct animal life. Because of the limestone rock, this is a good place to find typical lime-loving plants such as mountain avens and holly fern.

THE FLOW COUNTRY *Highland*

Although the scientific interest of the numerous bogs lying across the top end of Scotland has been recognised for some time, it was really not until the threat of extensive conifer afforestation emerged in the mid-1980s that the true value of these wetlands was fully appreciated. In marked contrast to the more obviously spectacular mountain and coastal scenery of much of north-west Scotland, the relatively flat rolling blanket bogs of north-east Caithness and east Sutherland may seem monotonous, somewhat akin to the vast tundra of the Arctic, treeless and desolate. This is however the largest continuous expanse of peat moorland in Britain, now acknowledged to be a scarce ecosystem in world terms and therefore of international importance for conservation.

This complex system dominated by a deep mantle of peat with literally thousands open water pools, as well as larger lochs, supports a highly specialised flora and fauna with strong affinities with comparable arctic and subarctic wetlands. The surface is mostly a spongy, living layer of sphagnum moss which is continually building upwards. Below this, the peat, with its preserved pollen grains, represents an archive of the plant history of the region, recording its development and environment over thousands of years. The whole area is the summer breeding ground of a fascinating bird fauna, typical of mainly northern species, but here inhabiting a southern and oceanic region. Spread over this large area, the number of several local moorland birds is a substantial proportion of their total European populations south of the Baltic, including particularly important breeding populations of golden plover, dunlin, greenshank and arctic skua. The lochs and smaller *dubh lochans* support breeding red-throated and black-throated divers, greylag geese, wigeon, teal, common scoters, and red-breasted mergansers. Rare breeding waders include Temminck's stint, ruff, wood sandpiper and red-necked phalarope. Raptors such as hen harriers, golden eagle, merlin, peregrine and short-eared owl also use the bogs as breeding or feeding areas.

FORSINARD

Lying in the heart of the 'Flow Country', this reserve offers an unusual opportunity to get to close quarters with a habitat which is otherwise difficult to penetrate. The visitor centre and associated nature trail allow the public to cross the dubh lochans and hear the various bird calls, from trilling dunlin to the ghostly call of divers on the pools. Hen harriers commonly hunt over the open bog and merlins are frequently seen in the heather on top of the larger rocks. On the fields around the centre, visitors may see small flocks of golden plovers, greylag geese, lapwings and curlews, while otters frequent the river. It is not unusual to see herds of red deer in the vicinity.

On the A897 14 miles from Helmsdale
L R WC I *(RSPB)*

INVERNAVER *Highland*

One of the most unusual reserves in demonstrating montane flora at very low altitude, with dwarf and prostrate juniper growing as a bush alongside heathers and coastal plants. The great sweep of lime-rich shell sand contrasts with the exposed moorland conditions above, where bearberry, mountain avens, and crowberry grow alongside sea campion and thrift! Here can be found Scottish primrose in the more open sandy areas, while elsewhere the distinctly montane yellow and purple saxifrage and alpine bistort thrive. Variety is added by the birchwood which grows along the River Naver, and animals which have been observed include otter, wildcat, badger, and fox. Birds of the woodland such as sparrowcock and woodcock occur, with predators such as kestrel and buzzard hunting across the more open moorland, where lochans provide a habitat for breeding waders.

Off the B871, half mile west of Bettyhill.

L *(SNH)*

Dunnet Head *(ND 2177)* The cliffs have many breeding seabirds such as puffins and black guillemots, and in the moorland behind the coast, great skuas, twites and ravens are frequently seen. To the west, in Dunnet Bay, there is a range of seaduck and divers in winter.

Dunnet Links *(SNH ND 2269)* The lime-rich sand dunes and grassland combined with wet areas, have produced a very rich flora including the Scottish primrose and a number of mountain plants growing here at very low level. The shelter of the plantations provides an excellent butterfly habitat.

LOCH FLEET *Highland*

Enclosing a large tidal basin with extensive mudflats, this reserve is a haven for shelduck, oyster catcher and redshank in summer, and later in the year, there are large numbers of curlew, golden plover and knot. It is an important resort for such wintering duck as wigeon, teal, golden eye, long-tailed duck and mallard. Common seals are frequently seen on the sandbanks from the minor road along the south shore. An unusual feature is the pine plantation with very typical pinewood flowering plants such as the delicate mauve twin flower, one-flowered wintergreen, and lesser twayblade. Sand dunes and coastal heath add to the great variety of habitats on this attractive coastal area. There are several lay-bys around the reserve from which the birdlife can be viewed and guided walks are provided in summer.

By the A9, 2 miles south of Golspie.

L I R *(SWT)*

Nigg and Udale Bays/Munlochy Bay *(Highland Region and SNH/RSPB)* These two reserves, noted for their intertidal birdlife, are on either side of the Black Isle promontory. They lie within the Cromarty Firth, of international importance for its wintering wildfowl and waders. Viewing of both areas is best from the roadsides On **Nigg** *(NH 790730)* and **Udale Bay,** *(NH 7165)* the speciality is the large number of wigeon, mute and whooper swans which feed on the plentiful eel grass. Many of the autumn greylag geese of this area, numbering in excess of 10,000, roost here, while the plentiful shellfish attract waders in their thousands. Viewing of both areas is best from the roadsides and from the new hide at Udale Bay. The sheltered **Munlochy,** *(NH 6753)* in addition to its many wigeon, supports around 200 shelduck, several hundred teal and provides a safe roost for greylag and pin-footed geese. The ornithological interest of Munlochy continues into March, whereas the peak period for Nigg and Udale bays is prior to December.

Ben Wyvis *(SNH NH 460680)* Noted for its great sheets of summit moss heath and lichen-rich blanket bog, as well as late snow-bed vegetation on this exposed mountain.

Falls of Shin *(NC 5899)* One of the best places to see Atlantic salmon (see *The Leaper*) leaping up these falls to the viewing site (including access for the disabled) between May and November.

LEDMORE AND MIGDALE WOODS *Highland*

Covering over 7,000 hectares, this is the largest single property of the Woodland Trust, with outstanding Scots pine and oakwoods of national importance. Located on the north shore of the Dornoch firth, the site includes a wide variety of habitats, apart from its woodland, such as valley bog, heather moorland, conifer plantations and rocky crags set in a wild and beautiful landscape, with spectacular views of the mountains of west Sutherland. Ledmore oakwood is the largest remnant of ancient oakwood in Sutherland and the most northerly oak wood of any extent in Eastern Britain. There is also a large area of naturally regenerating birchwood, while the wetlands include open water, reed beds, and scrub margins of willow, alder and birch. The bird life is accordingly very diverse, including wood and willow warblers, redstarts, crossbills, tawny owls and buzzards, while osprey are frequently seen fishing on nearby Loch Migdale. Red, roe and sika deer are all present. The site demonstrates a long history of human habitation from prehistoric chambered cairns to more recently ruined crofts. It has the particular interest of once being owned by the great Scots businessman and philanthropist, Andrew Carnegie.

At Spinningdale, 3 miles east of Bonar Bridge
L *(WT)*

North Kessock *(NH 655478 and NH 749557)* The car park approximately 1 mile from the north end of the Kessock Bridge overlooking the Beauly Firth has become a well-known viewing point for observing the only resident population of bottlenose dolphins in the North Sea and harbour porpoises can also be seen here. (Another good location is **Chanonry Point** further north near Fortrose).

Cromarty Firth *(NH 865757)* The firth has now become well known for viewing marine mammals, especially harbour porpoises and bottlenose dolphins. Seals are often seen at Seal Point (Foulis Point) from the A9 at *NH 598635* and over 100 regularly haul out on the opposite shore near Findon Mains *(NH 600610)*.

SKYE *Highland*

The largest of the Inner Hebridian islands, Skye is remarkable for its range of landscapes from the soft wooded Sleat Peninsula to the exposed peaks of the Cuillins and the dramatic scenery along the north-east coast around Trotternish. This last area demonstrates landforms found nowhere else in Scotland, with a spectacular scenery of landslips, leaving isolated basaltic columns, some of which, at the Quirang, have collapsed to create a landscape of rock pinnacles interspersed with moist green meadows and tiny lochans. On the coast, the 'kilt rock' formations have become tourist attractions, with waterfalls dropping sheer into the sea. By contrast, the jagged gabbro of the Black Cuillin, and smooth pink granite of the Red Cuillin combine their contrasting shapes to form a mountain area of dramatic and distinctive outlines of great scenic splendour. These remarkable hills dominate much of the island seaboard of north-west Scotland, so that they can be seen from places as far apart as Ardnamurchan and northern Lewis. Glen Brittle has magnificent waterfalls, while Torrin is dominated by the grey slabs of Bla Bheinn, and Elgol provides the classic view of the Black Cuillin in serried ranks above the concealed amphitheatre of Loch Coruisk.

These famous mountains fall within a National Scenic Area, and a large segment has been acquired by the John Muir Trust, including Torrin, Strathaird, and Sconser. Torrin has a wide range of habitats: limestone grassland, acid heath and moorland, rocky mountain summits, freshwater lochs, saltmarsh sandy and rocky shores. Adjoining is Strathaird, which extends into the heart of Cuillin at Loch Coruisk, including the crofting townships of Elgol, Glasnakille and Drinan on the Strathaird peninsula. At Sconser, the Trust property takes in one of Britain's highest woodlands – dwarf alpine juniper clinging to the rocks at over 800m. An interesting and challenging feature of the management of these extensive properties is the inclusion of crofting land which is still used for agriculture and where the local crofting community is very much involved in the planning for long-term conservation of both the land and its people.

(JMT)

Old Man of Storr *(NG 5154)* Superb panorama of eroded pinnacles and cliffs, with a variety of arctic-alpine plants growing at relatively low altitude, including the rare Iceland purslane.

Kylerhea Otter Sanctuary *(FC NG7822)* A forest nature reserve where closed circuit television allows visitors to watch otters in the wild – the only such facility in Scotland.

KINTAIL *Highland*

The peaks known as the Five Sisters of Kintail - four of which are 'Munros' (over 3,000ft) – form the core of this range rising from sea level at the head of Loch Duich, among the most sheer of grassy mountains in Scotland, with impressive corries, crags and precipices. Here there are very large herds of red deer and smaller numbers of wild goats. Nearby, the Falls of Glomach (see separate site description) at 123m are among the highest in Scotland, with an especially spectacular leap into space. The mountains support a varied and interesting arctic-alpine flora, such as trailing azalea, purple and mossy saxifrage, and alpine meadow rue.

The NTS countryside centre at Morvich, which is the best point of access to the mountains is open from May to September and is the base for the ranger service.

North of A87 between Lochs Chuanie and Duich, 16 miles east of Kyle of Lochalsh
L R WC I *(NTS)*

BALMACARA (including LOCHALSH WOODLAND GARDEN) *Highland*

A large crofting estate with magnificent views over Skye and Applecross, lying on the rugged promontory between Loch Alsh and Loch Carron, ranging from rocky shores and island to typical west coast hill ground, with many bare rocky knolls interspersed with peaty lochans and heathery slopes – the coastline is especially attractive with its many islets, frequented by otters. The several walking routes can be linked to the waymarked paths operated on the adjacent Forestry Commission property. Traditional crofting is still carried out from the townships of Duirinish, and Plockton, where visitors are asked to avoid trespassing. Lochalsh Woodland Garden has pleasant sheltered lochside walks with a variety of tree species including Scots pine, oak, beech, and a range of exotics including rhododendron and bamboos.

A87, 3 miles east of Kyle of Lochalsh
L R WC I *(NTS)*

Eilean na Creige Duibhe *(SWT NG 824335)* This rocky wooded island is close to the south shore of Loch Carron near Plockton, where a boat can be hired. The island is dominated by Scots pine, and has a well-established heronry, while eider nest in the heather above the shoreline. Otters frequently come ashore here.

ABERNETHY/LOCH GARTEN *Highland*

Renowned as the first osprey nesting site to be made accessible to the public, Loch Garten, in the heart of the Abernethy pinewoods has a well-equipped observation centre overlooking the main nest site, which attracts around 10,000 visitors each year. Closed circuit television allows visitors an enthralling close-up of the activities of the birds and their young. Ospreys have now bred here successfully for 28 years, but the surrounding forests and lochs are also notable for many other breeding bird species, including Scottish crossbills, capercaillie, black cock, golden eye, wigeon and teal. Greylag geese, whooper swan and goosanders roost on Loch Garten in winter. Crested tits make their nests in the dead pine stumps, and the forest is noted for its many old trees. Elsewhere, stunted bog pine reflect the conditions in the wet hollows which are a particular feature of this reserve, comparable to the Scandinavian forest bogs. This site is important for its invertebrate fauna, including no less than 350 beetle species, 11 dragonflies, and over 280 moths and butterflies, many of these insects being rare northern species. At Abernethy, the RSPB have carried out innovative conservation management to remove plantations and reduce red deer numbers to allow for active natural pine regeneration.

Signposted off Boat of Garten – Nethybridge Road B970
L R I * *(RSPB)*

CAIRNGORMS see main description under Aberdeen and Grampian

ROTHIEMURCHUS *Highland*

Part of the extensive Cairngorms National Nature Reserve, the native pinewoods have good facilities for visitors, many attracted to the very beautiful setting of Loch an Eilein, surrounded by fantastically shaped old pine trees, over a carpet of heather and juniper bushes. A winding nature trail makes a circuit of the loch and two visitor centres on the Rothiemurchus Estate provide information on the wildlife and land use of this varied countryside, with its farming, forestry, sporting and recreational interests. Glenmore Forest Park, in addition to its considerable wildlife interest, also offers many outdoor recreational pursuits, especially in the vicinity of Loch Morlich, with information being provided at the visitor centre operated by Forestry Commission.

1 mile east of Aviemore off A951
L R WC I *(Rothiemurchus Estate/SNH)*

Craigellachie *(SNH NH8812)* A fine open birchwood and moorland under the cliffs above Aviemore with a loop trail including an excellent viewpoint over the Spey Valley.

CREAG MEAGAIDH *Highland*

The stupendous ice-carved crags and cliffs of Coire Adair make this reserve a mecca for winter climbers and mountaineers, offering some classic climbs, with many of the springlines and wet faults freezing in winter to produce cascades of steep ice. The wildlife is equally outstanding: stretching from the heights of Creag Meagaidh itself at over 1,100m down to the shores of Loch Laggan, the National Nature Reserve exhibits a range of semi-natural vegetation surpassed by few other sites. In an area which has elsewhere been extensively afforested, Creag Meagaidh still has expanse of open moorland important for its several birds of prey, including golden eagle and such typical mountain birds as red grouse, ptarmigan and the rarer dotterel (see *Something to Grouse About*) – a young bird ringed here in 1987 was discovered overwintering on the north-west coast of Morocco!

Here many native tree species such as birch, alder, willow, rowan and even oak are now regenerating naturally because of reduced grazing. The moorland supports a variety of insects, from the pearl-bordered fritillary to the mountain ringlet, and the large golden ringed dragonfly can be seen hawking along the stream edges in summer. Tufted duck, golden eye, pochard and mallard are present on Loch Laggan and in summer the osprey is a frequent visitor; in winter the loch is used for feeding and roosting by whooper swans and cormorants.

There is a constructed path from Aberarder to the Coire Lochan in Coire Adair with its magnificent views of the ice-sculpted landscape above.

Off the A86, 7 miles south west of Newtonmore
L R *(SNH)*

Insh Marshes *(RSPB NH 775998)* Some of the most extensive river marshes in the whole of Britain, here around the Upper River Spey against the backdrop of the Cairngorms, with abundant wetland and woodland birds which can readily be observed from the viewing hides. Especially important for breeding and wintering wildfowl, the reserve also holds good numbers of breeding mallard, wigeon, teal and tufted duck, while the rare wood sandpiper feeds regularly here. Osprey and hen harriers are regularly seen hunting over an area whose diverse habitats provide a very impressive bird list.

Glen Tromie *(NN 782972)* Birch woodland with their redpolls, siskins and redstarts contrast with heather moorland above, where hen harrier, merlin and peregrine can be seen in this pleasant river valley near Kingussie.

CHAPTER 17

Outer Isles

THIS AREA INCLUDES BOTH the Western Isles (Outer Hebrides) and the Northern Isles taking in the Orkneys and Shetlands. Although these island groups are quite dissimilar in their geography, ecology and culture, they are grouped together here for convenience. With the exception of Harris in the Western Isles, the altitude of these islands is modest, but their exposure to the westerly winds means that they share a lack of tree and shrub cover. Their geology however is quite distinctive, with virtually all of the Western Isles being dominated by the ancient Precambrian rocks of the Lewissian series, while those of the Northern Isles are much younger: mainly Old Red Sandstone of Devonian age in the Orkneys, while the Shetlands have a mixture of this rock and also the older Dalradian series and even some granite in the north.

In that respect, Shetland has more affinity with the Western Isles, since these hard acid rocks tend to produce poorer peaty soils, distinguishing this 'black' (because of the peat) archipelago from the considerably more fertile soils of 'green' Orkney to the south. As a result, it is claimed with some justification that Orcadians are 'farmers who sometimes fish' while Shetlanders are 'fishermen who sometimes farm'! However these soil conditions can be considerably modified by superficial deposits, notably of blown and lime-rich shell sand to create the flowery dune grassland known as machair, and which for example, distinguishes much of the west coast from the extensive peaty moorlands of the hinterland. On Orkney where some arable cultivation is possible, full-time farm holdings may be of average size, but elsewhere crofting agriculture is usually part-time and on relatively small acreages.

The islands share transport problems, both with the mainland of Scotland and not least between individual island groups, such as between Orkney and Shetland, which makes for a high cost of living due to transhipment and costly air transport, only partly offset by subsistence crofting and the availability of peat for fuel. Although oil developments have considerably enhanced the economy of the Northern Isles and have greatly contributed to local services, their impact on the landscape and way of life has been far less than initially feared.

Another important difference affecting land holding and use is that the Northern Isles were not subject to land clearance in the 18th and 19th centuries to anything like the extent of highland Scotland, including parts of the Western Isles. Nor are the Northern Isles particularly suitable for gamesport, so that there has not been the establishment of dedicated deer forest and grouse moors are very limited. While relative to the rest of Britain the population of the Outer Islands would be described as 'sparse', the Northern Isles have not suffered the same level of emigration and drastic depopulation as the Western Isles, where the resident population has dropped to one third of what it was even a hundred years ago. A visitor would immediately notice strong cultural differences between these two island groups: whereas the Norse influence is still very strong in the Orkneys and Shetlands, especially in language, literature and festivals, the Western Isles, although also ruled by the Nordic 'Lords of the Isles', appear to have more affinities with a western Celtic culture, including the retention of the Gaelic language, which is not spoken in the Northern Isles.

All of the island groups are dominated by the sea, and the coast provides everything from great stretches of curving white sand beaches to forbidding cliffs, among the highest in Europe, thronged with seabird populations as large as any in the North Atlantic, and therefore of international importance – outstanding examples are **St Kilda, Noss, Fair Isle,** and **Noup Cliffs**. Of all the regions of Scotland, this is the one most renowned for the number and size of its breeding seabird colonies, particularly in Orkney and Shetland, where the RSPB administer a greater number of reserves than anywhere else in Britain - an ornithologists paradise. Machair already referred to is well represented at Balranald, while **Loch Druidibeg** and **Loch Bee** are good examples of the thousands of water bodies which are such a feature of the Western Isles. The machair and its associated environmentally-friendly cultivation under the traditional crofting system is now recognised as a unique form of land use which provides a wealth of wild life, both plants and especially ground-nesting birds, and where the intensive agricultural methods of the rest of lowland Britain would be inappropriate and uneconomic. Apart from coastal cliffs and wetlands, the moorlands of the island groups are important for a wide range of birds, including predatory species such as hen harrier, relatively unmolested by sporting interests.

The islands are rich in history, but Orkney is outstanding for its sheer density of important archaeological monuments, from the awe-inspiring burial chambers at Maeshowe to its massive stone circles at Stenness and Ring of Brodgar, rivalled only by the equally impressive Standing Stones of Callanais on Lewis. On the same island, on the west coast, the restored Black House and street in the old village of Arnol provides a glimpse of living conditions in a crofting community of relatively recent times. On Shetland, the extensive ruins at Jarlshof reflect layers of settlement over many centuries. The Broch at Mousa on Shetland is one of the best examples of these fascinating defensive structures.

The experience of visiting the Outer Isles is unique, with a landscape, culture and history found nowhere else and with wildlife riches which will remain long in the visitor's memory – rolling Atlantic breakers crashing on to uninhabited shores, desolate moorlands resounding only to the cry of birds, the grey jagged mountains of Harris and the lovely 'whaleback' islands of the green Orkneys. All of this is accessible to the visitor prepared to carry out some quite careful planning to overcome the complications of communications which are also very much at the mercy of wind and weather – so do not be in a hurry!

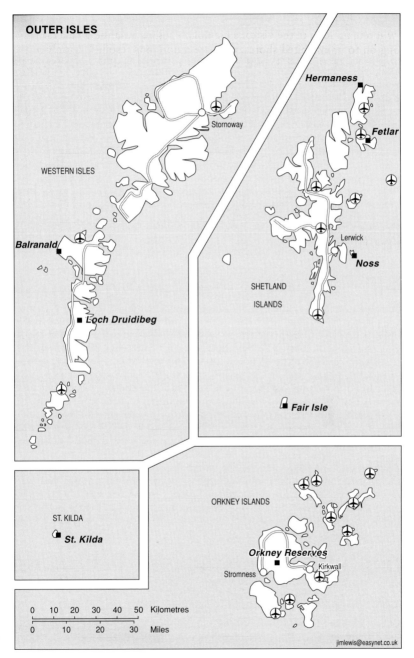

OUTER ISLES

Hermaness

Fetlar

Stornoway

WESTERN ISLES

Balranald

Lerwick

Noss

SHETLAND

ISLANDS

Loch Druidibeg

Fair Isle

ORKNEY ISLANDS

ST. KILDA

St. Kilda

Orkney Reserves

Kirkwall

Stromness

| 0 | 10 | 20 | 30 | 40 | 50 | Kilometres |

| 0 | 10 | 20 | 30 | Miles |

jimlewis@easynet.co.uk

LOCH DRUIDIBEG *Western Isles*

This is the home of one of the few surviving colonies of native greylag geese in Britain, resident all the year round, but the National Nature Reserve also has a remarkable range of other species, including red-breasted merganser, hen harrier, short-eared owl, buzzard, mute swan, merlin, snipe, redshank, dunlin and even the occasional golden eagle. Many of these species are attracted to the open water and surrounding wet peaty moorland, dotted with bog asphodel and cotton grass, while in the pools water lobelia is common. Its western sector provides a complete contrast in the lime rich machair and sand dunes, carpeted with flowers in season, with abundant orchids, while arctic tern and ringed plover nest in the unvegetated areas. The many islands in the main loch are unusual in supporting a quite dense tree and scrub cover of Scots pine, juniper, willow and rowan. The meadow land behind the machair is important for corncrake, corn bunting and twite.

In the northern part of South Uist between Stilligarry and Howmore
L R *(SNH)*

ST KILDA *Western Isles*

Permission to stay on this remote island must be obtained from the National Trust for Scotland, but any visitor who succeeds in making the journey to the only natural World Heritage Site in Scotland is likely to claim it as one of the experiences of a lifetime, especially for keen ornithologists. However, the remarkable history of the community, which survived here until the evacuation of the 1930s, casts its own spell, with much of the original village, burial ground and stone 'cleits' for storage of the harvested gannets still preserved. The landscape and birdlife of this archipelago are simply stupendous, and recently the waters around these islands have been found to be unusually rich in marine flora and fauna (see *The Marine World*). The north face of Conachair drops an astonishing 430m into the sea – the highest sea cliff in Britain. Amongst the multitude of birds, perhaps the gannet takes pride of place as the largest gannetry in the world at over 50,000 pairs. Apart from the multifarious birdlife, including large colonies of puffins, Manx shearwaters and Leach's petrel among the specialities, the islands host the St Kilda wren and the St Kilda field mouse, both distinct island sub-species, while the flocks of wild Soay sheep are still easily seen.

L R *(NTS/SNH)*

Loch Bee *(NF 7745)* Offers the fine sight from the A865, of several hundreds of mute swans on this loch which stretches virtually across South Uist.

BALRANALD *Western Isles*

On the island of North Uist, this reserve is known for its remarkable density of nesting waders – lapwing, ringed plover, oyster catcher, redshank and dunlin. Some 50 bird species nest annually on the reserve, including both arctic and little terns. The distinctive call of the elusive corncrake is heard here in one of its strongholds in the Western Isles. In early summer, the machair is a sheet of colourful flowers maintained by the crofting system, providing not only a delightful flora, but also an abundant insect fauna which helps to support the large numbers of meadow species such as skylarks, twite and corn bunting and which have become locally scarce on the mainland. Offshore, seals and occasionally porpoises, dolphins and whales are not uncommon – there is a breeding colony of grey seals on the island of Causamul – and otters are sometimes seen in the freshwater lochs.

What makes this reserve especially attractive to birds is the shallow Loch nam Feithean and its variety of fringing habitats, several of which are important for breeding duck, as well as the waders mentioned. These include gadwall, wigeon, tufted duck and shoveler. Migrant and wintering species such as scaup, common scoter, pintail, whooper swans and several species of geese use the reserve, and the marshes provide good hunting territory for merlin, hen harrier and occasional peregrine. The nearby Valley Strand is an excellent viewpoint from the road for watching waders feeding on the great tracts of intertidal mud.

Off the A 865, 20 miles west of Lochmaddy

R L WC I *(RSPB)*

Stornoway Woods *(Western Isles Council NB 419333)* These woodlands provide a rare sanctuary on the relatively treeless island of Lewis for a variety of woodland bird species, such as spotted flycatcher, treecreeper, and goldcrest together with several nesting ravens.

Gress to Tolsta *(NB 50/43)* From the B895 north of Stornoway, this is a good place to watch for breeding arctic and great skuas or 'bonxies'

Tiumpan Head *(NB 574377)* breeding ravens are a feature of this coastal headland to the north-east of Stornoway, which also provides good opportunities for watching seabirds.

FAIR ISLE *Shetland*

The bird observatory on this remote island has probably recorded more rare birds than any other in Britain, many of these from the Arctic. Caves, arches and stacks are a feature. Porpoises and whales are frequently sighted off-shore, and seals are commonplace around the sea caves, while the farmland and moorland supports over 200 species of plants. With 30,000 puffins and 25,000 pairs of fulmar, among many other seabirds, the island is truly a bird haven. Just as fascinating as its outstanding natural history is the thriving human community, surviving in the most isolated situation anywhere in Europe

Access by air and ferry: the island can provide accommodation for up to 50 persons - contact Shetland Islands Tourism.

L R WC I *(NTS)*

NOSS *Shetland*

Some of the most impressive seacliffs in Europe, nearly 200m high, are thronged with gannets, guillemots, fulmars, puffins and kittiwakes, while on the moorland, over 200 great skuas and 40 arctic skuas harry any intruding birds. Over 100,000 birds breed in this seabird city of Old Red Sandstone, while the moorland supports a wide variety of mountain and coastal plant species. Probably the most spectacular headland, the Noup, has 5,000 pairs of gannets, while below there are over 60,000 guillemots and 10,000 kitti-wakes – altogether an awe-inspiring sight. There is added historical interest in the Shetland pony 'pund' where mares were kept as breeding animals for the coal mines. Although an island, it is easily accessible by ferry service *(inflatable boat)* from the nearby island of Bressay from late May to end of August except Mondays and Thursdays.

L R WC I *(SNH)*

Loch of Spiggie *(RSPB HU 3716)* The most important freshwater site for wintering wildfowl in Shetland, but in summer skuas, kittiwakes, and arctic terns bathe in the loch, with breeding duck and waders on the shore.

Mousa *(HU 4624)* Famous for its ancient broch, this island is home to one of the few accessible storm petrel colonies in the country.

Unmistakable with its eye-spotted bands of red or orange on its very dark brown wings, the Scotch Argus butterfly is widely distributed in Scotland on moorlands over 90m.

Making their way up river to their traditional spawning grounds, the Atlantic salmon, sometimes known appropriately as 'The Leaper', is a noble sight surmounting the falls.

The showy yellow flowers of the flag iris form dense clumps
in wet ground throughout Scotland, but are especially
characteristic of marshy areas in the west.

The insect-eating round-leaved sundew makes itself as attractive as possible with its reddish-tinged and glistening leaf hairs to entrap insects of the boglands.

The acknowledged heraldic symbolic plant of Scotland, with its sharp spines, the spear thistle is only one of several species of this family.

Although a true mountain plant found up to altitudes of over 1280m, the attractive deep rose flowers of the moss campion can also be found on exposed coasts in the north and west down to sea level.

The great sheets of purple bluebells, or wild hyacinth, are a delight to the eye in open deciduous woodlands in spring.

The two commonest species of heather on the moorlands of
Scotland here growing together, provide an essential food
supply for grouse, and are maintained by periodic burning.

FETLAR *Shetland*

The speciality of this reserve is the red-necked phalarope, with 90% of the British breeding population, but there is also a wide range of breeding coastal and moorland birds, including red-throated divers, golden plovers, storm petrels, skuas and Manx shearwaters, to name only a few. The black guillemot or tystie is widespread. In spring, there are many passage migrants, including several rarities. Otters occur, and the island supports a large population of breeding grey seals, in addition to smaller numbers of common seal. Many distinctly northern plant species are present.

Car ferry from Yell/Unst or by air from Inverness/Aberdeen.
L R *(RSPB)*

HERMANESS *Shetland*

Renowned for vast numbers of puffins, now numbering tens of thousands, this National Nature Reserve also supports large numbers of shags, kittiwakes, fulmars and auks, especially at the north end, which is reached by a well marked footpath from the road end. The moorland has nesting skuas (watch out for their dive-bombing tactics – a walking stick or similar held above the head is a good idea!), golden plover and dunlin and it is relatively easy to see breeding red-throated divers on the lochans. The breeding gannet colony has increased spectacularly in recent years to over 5,000 pairs.

Around the cliffs twite and Shetland wren are frequently seen. Offshore, otters and both species of seals are seen regularly – grey seal breed in the caves below the cliffs. Because of the exposure of this very northerly site, moss campion, normally found at high altitudes, is found here at sea level. There is a new visitor centre in the old lighthouse below the car park.

Access by car ferry from Yell or air from Lerwick, then the B9086 road to Burrafirth
L R *(SNH)*

Keen of Hamar *(SNH HP6409)* Serpentine bed-rock here provides the habitat for a range of mountain and maritime plants growing together, including the rare Shetland mouse-ear chickweed.

Lumbister *(RSPB HU 5097)* The peaty moorlands and lochs are especially important for breeding whimbrel and red-throated diver, and have a thriving population of otters.

ORKNEY RESERVES

There is a greater concentration of accessible reserves of ornithological interest in Orkney than anywhere else in the country, and here they are grouped together as they are all administered by the RSPB. The sites are often important for their seabird colonies or moorland birds, but **North Hoy** *(HY 223034)* combines both of these, with its breeding great and arctic skuas, red grouse, golden plover, dunlins, hen harriers, merlins, short-eared owls and many others, while a variety of seabirds nest on the cliffs, including nearby a colony of Manx shearwaters. Peregrine and raven also nest here. Slightly surprisingly, there is a large population of mountain hares. The famous sea stack, the Old Man of Hoy rises to 140m – the highest in Britain, while the cliffs themselves are equally impressive at over 300m.

Noup Cliffs *(HY 3950)* on Westray are renowned for having one of the largest seabird colonies in Britain, with over 44,000 guillemots and nearly 13,000 pairs of kittiwakes. In the north west corner of the Mainland, **Marwick Head** *(HY 2224)* has smaller numbers of these species, but in the bay below, eider and several species of waders can be seen. The cliff top has spectacular displays of sea campion, thrift, and spring squill. By contrast, **The Loons** *(HY 245242)* is mainly wetland – one of the best remaining marshes in the Orkneys – which is not only important for its breeding wildfowl and waders, but has a rich flora, including grass of Parnassus, bog pimpernel, and knotted pearlwort. In the winter the flooded marsh attracts white-fronted geese and many duck. In that respect, it is similar to **Mill Dam** *(HY 481177)* on Shapinsay where in winter whooper swans and greylag geese can also be seen.

Egilsay, a recently acquired reserve on the island of that name, has the special interest of its breeding corncrakes, now under a special conservation management scheme. It is only a short distance westwards to the island of Rousay, where **Trumland** *(HY 427276)* supports dry heather moorland interspersed with wetlands which provide sanctuary for red-throated divers, and herring and lesser black-backed gulls, with golden plover and hen harriers nesting on the moorland – the latter preying on the local Orkney vole which occurs here. The great sweeps of **Birsay Moors** *(HY 3719)* represents a habitat which has declined substantially in these islands in recent years. It is noted for dense populations of short-eared owls and hen harriers, together with the unique ground-nesting kestrels, all which feed on the large numbers of Orkney vole infesting the grasslands. There is however a greater range of habitats to be found at **Hobbister** *(HY 3806)* near Scapa Flow, where the moorland is surrounded by a varied coast of sandflats, salt marsh and sea cliffs. It has a considerable range of breeding birds, including eider, shelduck, raven, rock dove, hen harriers, kestrel, short-eared owl, and merlin. The freshwater lochan holds breeding red-throated diver, as well as teal and tufted duck. **Coppinsay** *(HY 6001)* not only has huge colonies of breeding seabirds, but also a fine suite of coastal cliffs, and stacks providing a habitat for luxuriant cliff vegetation.

CHAPTER 18

Grampian and North East Coast

THE NORTH EAST CORNER of mainland Scotland is a region of considerable contrasts, embracing as it does everything from the sub-arctic wilderness of the eastern Cairngorms to the rich lowland farmlands of Buchan. In between there lies the great sweeping heather moorlands of upper Donside and a coastline which ranges from some of the largest sand dune systems in Britain to dramatic rocky cliffs. With the exception of the sea lochs and islands of the west coast, every other major Scottish landscape is represented, from high mountain to estuary. With a population of about half a million over a considerable land area, approximately half of this within the city of Aberdeen, this is far from being a densely populated area, reflected in extensive areas of varied and unspoiled countryside.

The western boundary is dominated by the massifs of the southern Grampians, forming an arc from which arise the major river systems of the Spey flowing northwards into the Moray Firth, and the Dee and the Don, flowing eastwards around Aberdeen into the North Sea. The Spey, with its source high in the Cairngorms, is probably the most active river system in Britain, carving its way in its upper reaches through hard granites and quartzites and, by its winter flooding, creating some of the most extensive natural marshlands in the country. The less turbulent Dee is especially renowned for its very particular surrounding landscape of mixed farming, forestry and sporting estates, which combined with its castles and association with royalty, have given Deeside a special attraction. All of these river systems are noted for their fine salmon fisheries and the quality of their waters on which the reputation of a number of internationally marketed malt whiskies are dependent.

The crescent of uplands on the western margins of the margin of the region lead down to fertile lowlands, with the area between the Don and the Spey claiming to be the largest continuous tract of agricultural land in the country, much of it reclaimed from peat bog in the 18th and 19th centuries. South of the Dee lies another important farming area behind the North Sea and stretching into North Angus, the windswept Howe of the Mearns, celebrated as the background to the novels of Lewis Grassic Gibbon. By contrast, large areas in Upper Donside are dedicated to game-sport, both deer stalking and grouse shooting, as well as upland farming. Notwithstanding the developments associated with oil exploration and production, based mainly on Aberdeen, this city and Peterhead to the north, are still important fish landing centres, although the smaller fishing villages round the coast have declined in importance. A feature of this region are the many planned towns and villages often characterised by their broad main thoroughfares, created for the encouragement of agriculture and industry in Tomintoul, Fochabers, Ballater, and Grantown.

The North Sea coast is a cold and exposed one, often with grey sea mists, or *haar* in the summer months, but towards the inner Moray Firth, the climate becomes considerably milder, with high sunshine records. Just south of Stonehaven, near to the most northern point of the great Highland Boundary Fault, the RSPB reserve of **Fowlsheugh** has one of the largest seabird colonies on the east coast of Scotland, but there are comparable and spectacular sea cliffs and seabird nesting sites at the Bullers of Buchan and Longhaven Cliffs. Among several sand dune systems, those at the mouth of the Rivers Ythan (**Sands of Forvie**) and Findhorn (**Culbin**) are especially notable in their extent and natural history interest. On the very southern boundary of the region **St. Cyrus** (described under Perthshire, Angus and Fife) combines a number of coastal systems on the margin of the estuary of the River Esk, with both sand dunes and cliffs.

Inland the valley of the Dee has a number of quite extensive remnants of the old Caledonian Scots pine forest, notably in **Glen Tanar** and in the Forest of Mar on the lower slopes of the eastern **Cairngorms**, both of these now managed as National Nature Reserves, with particular emphasis on securing natural regeneration of the native pinewood. Very few deciduous woodlands now remain but the managed policy woodlands around castles and great houses, such as at **Crathes** and **Drum**, albeit small, have a considerable number of plant and bird species. A more natural example of a previously more widespread native woodland is to be found in **Dinnet Oakwood** and in some of the nearby birchwoods.

The eastern mountain and moorland systems are well represented, with the Cairngorm tops being the largest area of relatively undisturbed natural vegetation above the treeline in Britain, and where some pockets of snow in north-facing corries are permanent. This and surrounding mountains such as **Lochnagar** provide important refuges for scarce arctic-alpine plant and montane animal species. (It is the high barrier of the Cairngorm range which helps to maintain the relatively dry climate of this region). Despite considerable losses in recent years of heather moorland to forestry and agricultural expansion, this semi-natural system is still extensive and is especially well demonstrated at the **Muir of Dinnet** in Deeside.

Apart from its considerable natural attractions, the region has a wealth of historical and architectural riches in the number of fortified houses, castles and mansions in addition to many more humble vernacular buildings. Deeside and Donside are especially known for their density of castles, many with preserved interiors and fine collections. Being one of the centres of Pictdom, there are plentiful relicts of that era in sculptured stones and the subsequent ecclesiastical history is exemplified by such outstanding structures as Elgin Cathedral and St. Machar's Cathedral in Old Aberdeen.

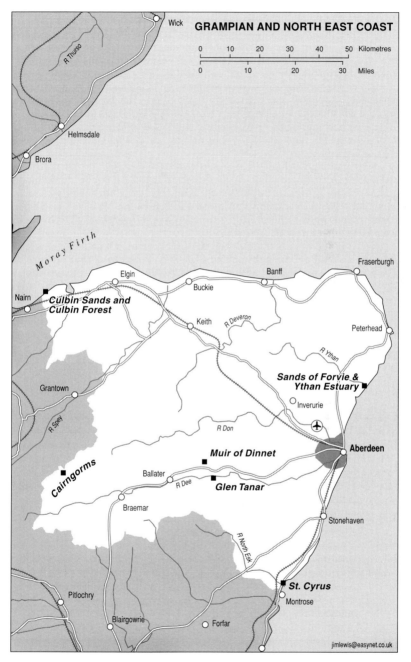

GRAMPIAN AND NORTH EAST COAST

| 0 | 10 | 20 | 30 | 40 | 50 | Kilometres |

| 0 | | 10 | 20 | | 30 | Miles |

Wick

Helmsdale

Brora

Moray Firth

Elgin

Banff

Fraserburgh

Nairn

Buckie

Culbin Sands and Culbin Forest

Keith

R Deveron

Peterhead

R Ythan

Sands of Forvie & Ythan Estuary

Grantown

Inverurie

R Spey

R Don

Muir of Dinnet

Cairngorms

Ballater

R Dee

Glen Tanar

Aberdeen

Braemar

Stonehaven

R North Esk

Pitlochry

St. Cyrus

Blairgowrie

Forfar

Montrose

jimlewis@easynet.co.uk

CAIRNGORMS *Aberdeenshire and Highland*

The largest nature reserve in Europe straddles two highland regions, but is placed here because the eastern sector is probably less disturbed by visitors than in Strathspey, where skiing facilities and easy access by chairlift are available. The Cairngorms justify the many superlatives used to describe its mountain fastnesses and high arctic conditions. The great granite massif reaches an altitude of nearly 1,311m between the Dee and Spey valleys, linked by the spectacular pass of the Lairg Ghru, where glacial action has carved out many high corries, several of these containing arctic-alpine lochs. The summit plateaux represent the largest area of high ground in the country, while the dramatic cliffs and extensive scree slopes lead down to tracts of heather moorland and some of the finest examples of the old Caledonian pinewood to be found anywhere in Scotland (see **Rothiemurchus** under Highland) The recent acquisition of Mar Lodge by the NTS will enable the eastern pinewoods to be protected from long-term overgrazing by deer and eventual full restoration.

Most of the flora and fauna typical of the central mountains of Scotland are represented here, from the golden eagle and ptarmigan of the high tops to the capercaillie and blackcock of the lower pinewoods. Here too are found the specialities of the pinewoods - the many large nests of wood ants (see *Antics in the Woods*) and the small pinewood orchids, creeping lady's tresses and lesser twayblade. Crested tit and crossbill are very much associated with these pinewoods, which are varied by birch, aspen, and an understorey of juniper in a variety of shapes, from columnar to almost prostrate. Red squirrel are relatively common and badgers are present. This is one of the very few places in Britain where a natural tree line has been able to develop.

Where outcrops of the richer schistose rocks occur, montane species such as roseroot, mountain sorrel, purple saxifrage, and sawort appear, as well as the much rarer alpine hair grass, starwort mouse-ear, and mountain hawkweeds. There are still considerable tracts of windswept heather moorland and grassland, with patches of mountain crowberry, cloudberry, bearberry, and bilberry at higher altitudes, interspersed with dwarf birch. While the sheltered pinewoods present no hazards, visitors intending to go into the mountain areas proper must be competent and well-equipped. The main visitor facilities and information services indicated below are based in Speyside (Coylumbridge) but are also available at Braemar on the eastern side.

L R WC I *(SNH/NTS)*

Morrone Birkwood *(SNH NO 1390)* A high level example of a sub-alpine birchwood with plentiful juniper and a very rich ground flora unusual under woodland conditions.

MUIR OF DINNET *Aberdeenshire*

Middle Deeside has a unique combination of landscapes from open moorland to oakwoods, pine forest, lochs and farmland, forming an intimated mosaic around the River Dee, itself one of the great river systems of Scotland, renowned for its salmon fishing. These diverse habitats are well represented by this reserve which is especially attractive in spring, with the colour of the young birches, and in late summer, the purple heather blooms over extensive moorland. Many visitors make for the spectacular rock cauldron of the Burn o' Vat, a huge pothole, which is part of a notable suite of glacial features, including Lochs Davan and Kinord, hoping for a sight of the family parties of otters which are regularly seen. The variety of habitats from open moorland to loch shore and pine and birchwood support no less than 140 bird species of which 76 breed regularly, and the reserve is noted for its diverse insect life, including many butterflies and moths such as the rare Kentish glory.

At the highest levels, the heather and bearberry heathland has been invaded by Scots pine and birch with the reduction in grazing stock over the last 25 years and since muirburning has stopped. Most of the woodland in the lower part of the reserve is birch, but there are smaller areas of mixed scrub of aspen, ash, hazel, and blackthorn around new Kinord. A feature of the moorland, which provides fine views along mid-Deeside, is the rich heath with abundant mosses and lichens, while many other species of bryophytes enjoy the damp shady conditions of the vertical walls surrounding the Burn o' Vat. The moor is also home to breeding black grouse and wintering hen harrier and merlin. The central lochs are especially attractive in autumn and winter, with considerable numbers of duck, whooper swans and large flocks of both greylag and pink-footed geese. Apart from its wealth of natural history, the reserve has considerable archaeological interest, particularly in its extensive Iron Age remains.

Off the A93, 8 miles east of Ballater

L R WC I *(SNH)*

Dinnet Oakwood *(SNH NO 464980)* A small northern oakwood noted for woodland birds such as great spotted woodpecker, pied flycatcher and wood warbler in addition to a varied ground flora, including chickweed wintergreen and common wintergreen.

Crathes *(NTS NO 7396)* A variety of mature woodlands of different deciduous and conifer species in the grounds surrounding the castle, with red squirrel and roe deer frequently seen, in addition to many small bird species which utilise these diverse habitats.

GLENTANAR *Aberdeenshire*

One of the delights of this easterly remnant of the old Caledonian forest are the trails along the banks of the River Tanar, a typical highland torrent with fine old mature pines and extensive natural regeneration on the heather-clad moorland above the main forest. In addition to pine, juniper is widespread, with more local birch, rowan, and aspen. Many of the attractive pinewood plants occur here, such as twinflower and lesser twayblade, as well as such characteristic birds as capercaillie, siskin and crossbill. There is a good chance of seeing several of the species of birds of prey. Otter, red squirrel, and wild-cat are present in this very fine example of the dry eastern type of the old pine forest of Scotland. A feature of the reserve is the active pinewood restoration management which is resulting in prolific regeneration of trees within the moorland.

Off the B976, 1 mile south-east of Aboyne
L R WC I *(SNH)*

Glen Muick and Lochnagar *(Balmoral Estate NO 2585)* Spectacular mountain and loch scenery with a variety of hill walks, with good opportunities for seeing red deer and golden eagle. The shores of Loch Muick have plentiful crowberry, bilberry, bearberry and cowberry, and in the wetter areas can be found the insectivorous sundew and butterwort. There is a good range of arctic-alpine species, including trailing azalea and least willow, as well as upland birds such as breeding ptarmigan and golden plover. Around the great granite blocks on the higher slopes, parsley fern and alpine lady fern are typical.

Bennachie *(FC NJ6821)* Around the conifer plantation there are labelled tree species; the area is known for its variety of non-flowering plants – forest walks lead to excellent viewpoints.

Drum *(NTS NJ 7900)* Superb oak, Scots pine and beech in this old open woodland, over luxuriant swathes of grasses and ferns, which also has ancient yews and wild cherries – a delight in spring, with the colour of the planted rhododendrons. Redpoll, garden warbler, great spotted woodpecker and blackpoll add variety to the birdlife.

SANDS OF FORVIE AND YTHAN ESTUARY *Aberdeenshire*

These high mobile dunes on the Aberdeenshire coast are the nearest thing to the Sahara in Britain, of interest in their own right, but also providing sanctuary for four species of terns and an important shelduck population. The National Nature Reserve is outstanding for its breeding eider duck – the largest concentration in Great Britain. In the autumn, the estuary attracts large numbers of waders, including uncommon species such as greenshank, spotted redshank, little stint and green sandpiper. In winter there may be as many as 10,000 pink-footed geese, with a variety of other wildfowl species. One of the most attractive habitats is the coastal heathland, found mainly in the north of the reserve. Here prolific crowberry and lichens grow, with creeping willow in the damper patches. Birch, pine and willow are slowly invading these heaths, particularly in the drier areas which support heathers and the yellow tormentil. In total, some 350 species of flowering plants have been recorded on this remarkable reserve – one of the least disturbed sand dune systems outside north west Scotland, which also has several nationally important archaeological sites. Although there is unlimited access to the dunes, visitors are required to stay on the footpaths in the breeding season between April and August. There is an observation hide overlooking the upper estuary which is easily reached by car and is available all year round. Visitors are especially encouraged to visit the new Forvie Centre and wildlife demonstration areas.

12 miles north of Aberdeen off the A975 between Newburgh and Collieston
L R WC I *(SNH)*

Loch of Strathbeg *(RSPB NK 063564)* A wetland excellent for viewing pink-footed geese and whooper swans in winter, while in summer, many breeding wading birds, and nesting sandwich and common terns can be seen. The number of wildfowl on this large shallow loch immediately behind the coast can rise to 20,000.

Aden Country Park *(Aberdeenshire Council NJ 984480)* Apart from its award-winning visitor centre illustrating agricultural life in this district, the park has some very old mixed woodlands, a varied riverside flora, and is noted for butterflies.

Fowlsheugh *(RSPB NO 876805)* The magnificent cliffs support no less than 80,000 pairs of six different species of seabirds – one of the largest accessible seabird colonies on the mainland of eastern Scotland.

CULBIN SANDS & CULBIN FOREST

Lying in both Highland region and Moray, and enjoying the distinctly mild climate of the Moray Firth, the area of sandflats, saltmarsh, and shingle bars managed as an RSPB reserve, is especially known as a wintering area for wildfowl and waders. Greylag geese frequent the shingle bars as a winter roost, while long-tailed duck, common scoter, and velvet scoter congregate in their thousands to seaward. The mudflats and saltmarsh provide a rich feeding ground for wigeon, mallard, and shelduck, and both curlew and bar-tailed godwit use the same area in large numbers. Predators such as peregrine, hen harrier and merlin are relatively common. In late summer, the shingle bars also provide a nesting area for several species of terns. Apart from its ornithological interest, the shingle bars represent an unusual physical coastal feature and support their own specialised flora.

In the Culbin Forest backing the open coast, the shingle is replaced by an extensive series of sand dunes, now thickly planted – mainly with Scots and Corsican Pine – as a dune erosion control measure, but birch and willow have survived and spread, especially in the wetter hollows. Among the heather and dune species can be found native woodland species such as chickweed wintergreen, creeping lady's tresses, coralroot orchid, and the lichen flora is especially notable. The forest is a haven for many animals and birds, including plentiful roe deer, many badgers, red squirrel, pine marten and wild cat. Breeding bird species include short-eared and long-eared owl, sparrowhawk, buzzard, great spotted woodpecker and water rail. The forest is a stronghold for breeding crested tits, with around 200 pairs being recorded in recent years.

Off the A96, 15 miles east of Nairn
L R WC *(RSPB/FC)*

Spey Bay *(SWT NJ 325657)* The finest active shingle ridges in Scotland at the mouth of this famous river with a rich assemblage of plants and invertebrates.

Speyside Way *(Moray Council NJ 349654 – 167367)* One of the 3 long-distance trails in Scotland is a 68km route following the lower Spey Valley from Tomintoul under the Cairngorms to the sea at Spey Bay (see above). Much of the route is relatively level along the old Speyside railway line through varied countryside with fine river views. The path frequently passes through attractive deciduous and conifer woodland, with many different flowering plant species and mosses and ferns, as well as a variety of butterflies.

St Cyrus: *Although lying within Aberdeenshire, for convenience and location, this has been entered under Angus below.*

CHAPTER 19

Perthshire, Angus and Fife

WITH THE KINGDOM OF FIFE as an outlier into the North Sea, the remainder of this varied region is dominated by the Tay, with the largest catchment of any river system in Britain. It rises far to the west before entering Loch Tay, itself one of several large deep highland lochs, such as Loch Earn and Loch Rannoch, which contribute so much to the attractiveness of this west Perthshire landscape. The whole of the region is backed to the north and west by the southern fringes of the Grampian mountain range, including the group known as the Breadalbane Mountains, which takes in **Ben Lawers**. From here a beautiful series of glens – **Glen Lyon**, Glen Garry, and Glen Almond contribute to the broad valley of Strathearn, while to the east, no less than five glens, from Glenshee to Glen Esk, cut through the foothills of Angus to open out into the mainly arable valley of Strathmore. From the mountains in the west to the Fife coast there is a significant drop in rainfall, while hill masses such as Ben Lawers experience some of the coldest temperatures in Scotland.

The Highland Boundary Fault (especially well displayed in the spectacular waterfall of Reekie Linn in Glenisla) cuts right across this region with the northern uplands being comprised mainly of metamorphic schists, while to the south, Devonian sandstones predominate. A band of limestone along the Breadalbane Hills, produces richer substrates. Volcanic outcrops create the low hills of the Sidlaws north of Dundee and the Ochils on the western border of Fife, where there has been extensive volcanic activity. While the typical effects of glaciation are everywhere to be seen, especially in the U-shaped glens, the line of small lochs along the southern margin of the hills – **Loch of the Lowes, Loch Kinnordy, Balgavies Loch,** Loch of Lintrathen – represent especially good examples of glacial depressions.

Fife, which is bracketed by the estuaries of the Tay and the Forth, is largely agricultural lowland, with a history of coalmining in the west of the county. However, the coast, especially along the peninsula known as the East Neuk, has a number of delightful small fishing villages such as Crail and St Monans, while the ancient ecclesiastical centre of St. Andrews is a magnet for visitors. Parts of its coast are outstanding for their natural history interest and the offshore **Isle of May** is renowned as a bird sanctuary. North of the great sand dunes and conifer plantations of **Tentsmuir**, the **Tay Estuary** is internationally important for its wildfowl populations. Further up this river, on the north shore near Errol, some of the most extensive reed beds in the country are to be found, now managed systematically for the production of thatching material. Here, along the Carse of Gowrie and the nearby Strathmore is some of the richest land in Scotland which, with its south-facing aspect, is especially suited to soft-fruit growing.

With increase in altitude, lowland arable agriculture gives way to upland grazings, forestry, and gamesport. Both the latter occupy extensive tracts, especially in the Angus glens, which still support large shooting estates. This dry heather moorland in the east is replaced by the wet moor and peatland of **Rannoch Moor** (see Highland) in the far north west of the region. Above the moorland, particularly where limestone outcrops, such as on Schiehallion and Ben Lawers, cliffs and riversides provide a habitat for many typically arctic-alpine plant species, while the mountain ranges are home to large herds of red deer, wild cat, blue hare, golden eagle, ptarmigan and many other species of the higher altitudes. At lower altitudes, the same limestone outcrops produce the herb-rich meadowlands which are a distinctive feature of upper Tayside, with their attractive floral displays from mid-spring onwards.

Perthshire has some good examples of native deciduous woodland, especially in the Tay Valley, at **Killiecrankie** and the **Birks of Aberfeldy**, although there has been much amenity planting with introduced species. Some of the earliest conifer plantings using species such as larch and Douglas fir are to be found within the Vale of Athol, notably on the banks of the Tay at Dunkeld and in the grounds of Blair Castle. An outstanding example of old native pinewood, with fine specimens of ancient spreading 'granny' pines, is found at the **Black Wood of Rannoch** on the southern shores of Loch Rannoch.

The lowland lochs and surrounding agricultural land and estuaries attract huge flocks of wintering wildfowl, **Loch Leven** near Kinross being perhaps the most notable sanctuary. The great congregations of greylag and pink-footed geese can be seen from late autumn to spring especially in the large arable fields of Strathearn and Strathmore. Ornithological interest is also a feature of the coastal of Angus, where the seabird colonies of Seaton Cliffs at Arbroath are an attraction, while further north **Montrose Basin** is renowned for its sheer variety of easily viewed waterfowl and waders.

The only major city in this region is Dundee, historically associated with the whaling industry, which together with the story Antarctic exploration, is well exhibited at Discovery Point near the harbour. Apart from this city and the previous coalmining towns of west Fife, there has been little industrialisation in the area, so that there is much attractive countryside, from well-kept farmland to highland loch and mountain. Not surprisingly, Perthshire especially has had a long standing-tourist economy, traditionally based on such resort towns as Crieff and Pitlochry. The last is a good place for visitors to see close at hand migrating salmon using the fish ladder constructed alongside the hydro-electric power station. Small towns within easy reach of the ski slopes at Glenshee, such as Blairgowrie, have been given a boost by this latest tourist development. Apart from the well-known golf courses at St Andrews, there are many other coastal 'links' which maintain large open green areas elsewhere, while the country parks at Monikie, Clatto, Camperdown and Crombie have considerable natural history interest. Added attractions in this region are the historic burghs such as Dunkeld and Arbroath with their early cathedrals and the Royal residences at Falkland and Scone.

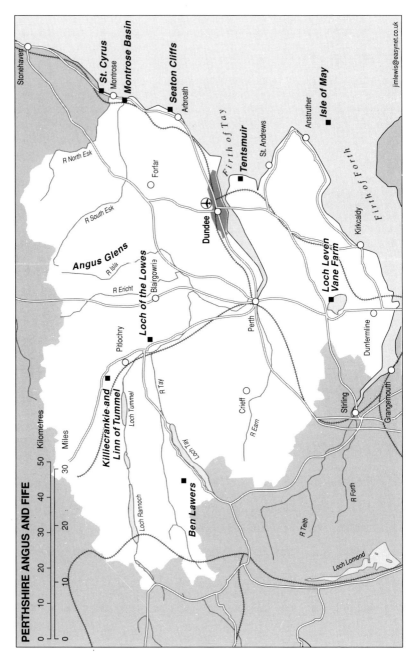

PERTHSHIRE ANGUS AND FIFE

jimlewis@easynet.co.uk

Stonehaven

St. Cyrus
Montrose
Montrose Basin

Seaton Cliffs
Arbroath

R North Esk

Forfar

Firth of Tay

Anstruther

Isle of May

St. Andrews

R South Esk

Tentsmuir

Dundee

Firth of Forth

Angus Glens

Kirkcaldy

R Isla

Loch of the Lowes

Blairgowrie

Loch Leven
Vane Farm

R Ericht

Pitlochry

Perth

Dunfermline

Killiecrankie and
Linn of Tummel

Loch Tummel

R Tay

Crieff

R Earn

Stirling

Grangemouth

Kilometres

Miles

Loch Tay

Ben Lawers

Loch Rannoch

R Teith

R Forth

Loch Lomond

50
30

40

20

30

20

10

10

0
0

BEN LAWERS

Situated above Loch Tay almost equi-distant between the east and west coasts of Scotland, this upland massif experiences a continental climate with low winter temperatures, and usually substantial snow lie. Together with its altitude and the presence of the Breadalbane limestone outcrops, it is therefore a particularly favoured site for the survival of arctic-alpine plants, for which it is renowned. The schistose rocks are very friable and give rise to cliff faces and ledges which are not only wet, but also continuously eroding to provide the exposed mineral soils which many northern plants of this group require. Elsewhere, the sides of mountain springs and mobile scree slopes provide alternative habitats for other rare or localised species.

Among the species for which this National Nature Reserve is noted are the vividly blue alpine forget-me-not, alpine cinquefoil, alpine saw-wort, and alpine fleabane, and woodland plants such as oak fern and many others. The dwarf herb grassland above 1,000m has a most colourful display of moss campion, mountain pansy, alpine mouse-ear, mossy saxifrage and Scottish pearlwort. In terms of the number of species and varieties of lichens, this locality is considered to be the most important in Britain for this group of plants. In addition, there is a rich insect fauna, including the small mountain ringlet butterfly which is a local speciality. Others are the northern eggar, yellow underwing, emperor moth, and fox moth.

Because of the importance of this site, considerable efforts, including large exclosures to keep out both deer and sheep, have been established on different parts of the mountain. The NTS visitor centre is a good starting point, providing exhibits on the natural history, management and land use of the area.

Off A827, 6 miles north-east of Killin
L R I WC *(NTS/SNH)*

Keltneyburn *(SWT NN 767508)* 5 miles west of Aberfeldy, this reserve is in a steep-side gorge supporting a rich deciduous woodland of native trees and shrubs, and adjacent to it, a flower-rich meadowland.

Glen Lyon *(NN 738472)* This long and very scenic glen, a few miles west of Aberfeldy, has at its upper end a fine example of relict scattered native Scots pinewood at Meggernie, and at the lower end, in the churchyard at Fortingall, several ancient spreading yew trees reputed to be 3,000 years old.

ANGUS GLENS *Angus/Aberdeenshire*

A series of relatively little known glens radiating like the spokes of a wheel on the southern fringes of the Grampians, Glen Clova, Glen Prosen, Glen Isla and Glenshee are all distinctive, some supporting large sporting estates, while others are mainly pastoral or dominated by new forestry. They lead from the rich farmland of Strathmore up into the sub-arctic conditions of some of the most famous plant localities in Britain. The high plateau and steep corries of Caenlochan, for example, stretching from Glenshee on the border with Aberdeenshire to the head of Glen Clova in the east, is comparable to the Cairngorms in its alpine habitats, where the cliff ledges are noted for such species as alpine meadow rue, purple saxifrage, alpine bistort and many others. On the exposed summit plateaus, woolly fringe moss, sedges and cladonia lichens are dominant.

It is easy to get to a high altitude by using the A93 Blairgowrie to Braemar road, where 15 miles south of Braemar, the Glenshee ski centre car park at over 650m is a good starting point to explore the grassland immediately to the west of the road. Here it is possible to find such characteristic high peatland plants as cloudberry, and in the drier areas, common cow-wheat, chickweed wintergreen and mountain pansy. Yellow and starry saxifrage, Scots asphodel, and alpine meadow rue grow in the gravelly streamsides. By using the chairlift to reach the summit ridge, many other montane plants are accessible, including trailing azalea and crowberry.

ST CYRUS *Aberdeenshire*

Just over the northern border of this region, this National Nature Reserve has a magnificent stretch of sands and low cliff – with its varied range of habitats, it supports over 350 flowering plants and ferns, and no less than 13 species of butterfly and over 200 moths. The beach and lava cliffs are good demonstrations of the effect of ice and changes in sea levels, and on the sand and shingle at the south end, there is a fluctuating little tern colony. There is an active inshore salmon fishery on the reserve, using fixed nets, and so it is hardly surprising that both otters and grey seals are quite common. The lower cliffs are especially colourful in late spring with the tall yellow spikes of great mullein, the deep wine of marjoram and the paler maiden pink. Unusual species such as the white Nottingham catchfly and white stonecrops also find their home here. On the beach the purple flowering sea rocket is an early coloniser, and in the dunes, more colour is added by spring vetch, forget-me-nots and wall speedwells. The scrubland is good for such bird species as the grasshopper warbler, stonechat, whitethroat, and yellow hammer. There is a regular passage of migrants through the area, including skuas and shearwaters, and in winter, the wader populations are among the best on this coast.

Off A92, 8 miles north of Montrose
L R WC I* *(SNH)*

The beadlet sea anenome is only one of the huge variety of marine organisms which make up the richness of Scotland's marine treasury, now yielding many new discoveries.

The calcium rich sands of the lightly-cultivated machair of our western coasts
provide an astonishing floral display from late spring onwards.

The rocky coasts of Scotland, including the islands and off-shore stacks and skerries, are some of our most unspoiled natural resources, of international importance for their breeding seabirds.

The great wilderness of Rannoch Moor, framed by an amphitheatre of mountains, is studded with sedge-fringed peaty lochans and glacial deposits of rocks.

The spectacular basalt lava flows at the Quirang in Northern Skye have been grossly distorted into weird shapes by massive landslides to produce a landscape unique in Britain.

The ancient eroded sandstones of the Torridon Hills of Wester Ross create some of Scotland's most dramatic mountain landscapes and provide a magnet for climbers.

The red deer, originally a large forest animal, here at home
in its natural habitat, the native Caledonian pinewoods of
Scotland.

A living mat of Sphagnum moss, of which there are many different species, is a good indicator of the health of the extensive peat bogs of Scotland.

MONTROSE BASIN *Angus*

This unusual enclosed river basin, with its mudflats and creeks, provides a rich feeding ground for thousands of waders and wildfowl at all seasons of the year. From late summer to spring, vast numbers of birds are present, best seen at high tide from the hides provided. It is renowned for its populations of redshank, knot, oystercatcher, curlew and dunlin. Several thousand wigeon are often present, with smaller numbers of other duck, including pintail, wigeon, shelduck and eider. It is an important moulting and wintering area for mute swans and is nationally significant for its 12,000 pink-footed geese, while 2,000 greylag frequent the area.

Immediately to the east of Montrose
L R WC I * (SWT)

SEATON CLIFFS *Angus*

Sea caves and stacks are a feature of the spectacular red sandstone cliffs, interrupted by the thick woodland of Seaton Den. A wealth of flowering plants include sheets of yellow primroses and purple violet on the gentler slopes in spring, while sea campion and purple milk vetch make a show later in the season. Where richer soils occur, carline thistle and clustered bellflower are indicators of unusually lime rich conditions. Many insects take advantage of the varied flora and shelter provided by the ungrazed vegetation, including several butterfly species and the colourful six-spot burnet moth. In summer, arctic terns and eider are present offshore, fulmar and herring gull nest on the few suitable rock ledges, and the caves shelter the breeding sites of rock doves. The site is nationally important for wintering purple sandpiper. The cliff-top nature trail provides a series of excellent vantage points

On the coast immediately north of Arbroath
L (SWT)

Loch of Kinnordy *(RSPB NO 3653)* Summer is dominated by the thousands of black-headed gulls which nest here, while in winter it is a favourite roost of up to 5,000 greylag geese. Ospreys fish in summer and the rare black-necked grebe have nested.

Balgavies Loch *(SWT NO 534508)* Open water and dense surrounding wet woodland provide habitats for breeding waterfowl, such as mute swan, sedge warbler, reed bunting, great crested grebe and wintering pink-footed and greylag geese in great numbers.

ISLE OF MAY *Fife*

Regular boat excursions from Anstruther and Crail allow visitors to spend up to 3 hours on this National Nature Reserve, now a stronghold for puffins in the south-east of Scotland, with over 8,000 occupied burrows. Other seabirds are well represented with over 1,000 shag and several thousand pairs of kittiwake and guillemot, as well as an expanding fulmar colony and common terns. There is a thriving grey seal population. A special feature of the reserve is the frequent arrival, during migration times, of birds of passage, including large number of commoner species such as bramling, but also very rare ones including gyr falcon and Sabine's gull which have been recorded from the first bird observatory in Scotland established in the early 1930s – it is still an important scientific research site. There is considerable archaeological and historical interest associated with the Chapel of St. Adrian, and the first manned lighthouse in Scotland.

L R WC I *(SNH)*

TENTSMUIR *Fife*

This large attractive area of sand dunes and foreshore has extensive pine plantations to landward, and is rich in both plant and insect life, especially butterflies, such as grayling, dark green fritillary, ringlet and small copper. Among the 400 plant species are many characteristic of dunes, with great abundance of grass of Parnasssus, for example, and the localised coralroot orchid in the wetter hollows. Birds include 9,000 waders such as grey plover, sanderling, little stint, and oyster catcher. The offshore sandbanks are used regularly by both common and grey seals and during the winter as a roost for pink-footed and greylag geese. The large flocks of eider in the Firth of Tay make this the one site in Britain which is internationally important for this species.

Signposted off the B946, 10 miles south east of Tayport

L R WC *(FC/SNH)*

Morton Lochs *(SNH NO 4626)* Hides provide good viewing of the wide range of birds, from the warblers of the reedbeds to the migrant waders feeding on the mudbanks of these man-made lochs and breeding duck and mute swans.

Eden Estuary *(Fife Council NO 4819)* A local nature reserve known for its wildfowl and waders, especially the colourful shelduck, wigeon, oystercatcher, dunlin, redshank and knot.

LOCH LEVEN/VANE FARM *Perthshire and Kinross*

Framed by the Lomond Hills, the National Nature Reserve of Loch Leven is administered by SNH, while public viewing and educational facilities are provided by RSPB at Vane Farm on the south shore of the most extensive sheet of lowland freshwater in central Scotland, renowned for its brown trout fishery. The outstanding feature are the 23,000 pink-footed geese which arrive in the autumn, and their descent onto the loch, darkening an evening sky, is an awesome sight. There are also a substantial number of greylag geese, whooper swans, wigeon, teal, mallard and pintail – the reserve holds the greatest concentration of breeding duck in Britain, mainly on St Serf's Island (where Mary Queen of Scots was imprisoned for a time in Loch Leven Castle) and is one of the most important wildfowl sites in Europe. Other goose species which are present most winters include a few barnacle, brent, white-fronted and snow geese. The wetter fields around the loch hold large numbers of breeding waders, including lapwing, snipe, and redshank. A nature trail guides visitors to a spectacular view of the loch above Vane Farm, where in summer, willow warblers, tree pipits, and spotted flycatchers are frequently seen, and in winter, redpoll and siskins. The fine RSPB visitor centre is one of the most popular in Scotland and is known for its educational programmes and varied viewing facilities.

Access to Vane Farm is signposted off the B9097, 1 mile east of Junction 5 on the M90

L R WC I * *(RSPB/SNH)*

Dollar Glen *(NTS NS 9699)* Tucked into the Ochil Hills, this dark steep-sided double gorge has been eroded through the ancient rocks by the Burn of Sorrow and the Burn of Care. Spectacular waterfalls and rapids lead to the dramatic situation of Castle Campbell above, through old deciduous woodland known for its spotted flycatchers, tree pipit, and woodpeckers, while mosses and ferns are abundant, alongside golden saxifrage, and on the drier slopes, wild garlic, dog's mercury, celandine and wood sorrel.

Cameron Reservoir *(Fife Council NO 470112)* Known as the most important pink-footed goose roost in the north-east of Fife, the reservoir also holds a variety of other wildfowl such as teal, goldeneye, wigeon, mallard, and greylag goose.

Dura Den *(WT NO 415145)* A well-known roost for large numbers of several bat species, the rich sandstone woodland has a varied flora and typical fauna.

KILLIECRANKIE AND LINN OF TUMMEL *Perthshire and Kinross*

The spectacular gorge and torrent of the Rivers Garry and Tummel provide the setting for these woodlands, pasture and moorland which have an abundance of wildlife. The rivers support otter and dipper, while the mixed oak, birch and alderwood are a haven for red squirrel and roe deer as well as many typical woodland birds, such as wood warbler, and the slopes above are known for buzzards, sparrowhawks and black grouse. Scotch argus butterflies are common in the grasslands. Both the green and spotted woodpeckers take advantage of the old woodland – Killiecrankie means the 'wood of the shimmering trees' – which are alive in early summer with redstarts, wood warblers and tree pipits. There are several species of orchids, in addition to grass of Parnassus and yellow mountain saxifrage. The river landscape is outstanding, and the deep pools, rushing torrents, splash ponds and calmer stretches of water are the homes of dragonflies, newts, otters, and goosanders, among many other highland river species.

Turn off main A9 road just north of Pitlochry onto B8079 to Killiecrankie.
L R WC I *(RSPB and NTS)*

Tay Forest Park *(FC NN 865597)* Set in renowned Perthshire scenery, this 16,000 hectare park offers everything from old Caledonian pinewood to beautiful lochs, attractive at any time of year, but perhaps most spectacular in its autumn colours. The wildlife is very varied, from the water birds of the loch systems, such as breeding red-throated divers to forest red squirrels and increasingly common pine marten – the capercaillie was re-introduced to Scotland here in 1837. Native Scottish crossbill frequent the pinewoods, and spotted flycatchers are typical summer birds of the deciduous woodlands. There is special conservation management for barn owls and goldeneye duck.

Black Wood of Rannoch *(FC/SNH NN 5755)* A very special relict of the old Caledonian pinewood on the shores of Loch Rannoch, noted also for its rare northern insects and moorland birds, including golden plover, dunlin, merlin, greenshank and red-throated divers.

Birks of Aberfeldy *(NN 855486)* A delightful woodland and river walk, thick with a variety of mosses and ferns, in addition to many northern plant species such as chickweed wintergreen, bird's nest orchid, common cow-wheat and alpine bistort.

LOCH OF THE LOWES *Perthshire and Kinross*

Set in varied scenery east of Dunkeld, the wildlife and surrounding landscape of this reed-fringed loch combine the features of both highlands and lowlands. It is especially well known for its breeding ospreys readily observed from the visitor centre. These magnificent birds first came to the loch in 1969 and have returned regularly ever since, arriving in late March or early April to refurbish their eyrie in an old Scots pine. The fascinating courtship displays of the great crested grebe are frequently seen on the reserve, home also to both little grebe and occasionally, the rare Slavonian grebe. There are many other kinds of waterfowl, including water rail, goosander, goldeneye and tufted duck, while terns fish here.

In winter, the resident ducks are joined by over 1,000 greylag and Canada geese. Also in winter, flocks of redpoll and siskin congregate in the alders and birches to feed. Redstart, garden warbler and tree creeper are among the summer bird visitors, and both green and great spotted woodpecker occur. Around the loch are attractively mixed woodlands containing at least 20 different kinds of trees and shrubs (including such native conifers as Scots pine and juniper), and there is a great diversity of water plants, including spectacular displays of both white and yellow water lily and bogbean. Other aquatic plants make a fine show, such as quillwort, and water lobelia.

2 miles north-east of Dunkeld by a minor road off the A923
L R WC I *(SWT)*

Stormont Loch *(SWT NO 193422)* Two shallow lochs with their fringing vegetation provide a habitat for wintering wildfowl, especially greylag, shoveler, and gadwall. Water plantain and yellow water lily are among the unusual plants found here.

The Hermitage *(NTS NO 0142)* A delightful woodland walk among some of the tallest Douglas firs in Britain and other fine specimens of early planted conifers. The river also supports a range of water birds such as dipper, grey wagtail, while willow warbler and long-tailed tit frequent the riverine woodland.

Tummel Shingle Islands *(SWT NN 9753)* On the banks of the River Tummel, with some 350 different species in the herb-rich grassland and other habitats, including alder, birch, and Scots pinewood, and with many varieties of both birds and insects (including several localised craneflies), this relatively small reserve is unusually diverse.

CHAPTER 20

Lothians and Borders

PROBABLY THE LONGEST SETTLED of all the regions of Scotland, this area stretches from the border with England to the Firth of Forth. Bounded on the east by the North Sea and on the west by a series of hill ranges reaching down as far as the Cheviot beyond the Tweed, the region has two main physical features – the fertile coastal plain to the north and east, and the rounded grassy and heathery hills and moorland of the Southern Uplands to the south and west, penetrated by the broad strip of lowland farmland known as The Merse along the banks of the River Tweed. This important river system starts high up in the rounded hills near the Devil's Beef Tub north of Moffat, before taking a winding course past famous border towns to debouch into the sea at the English town of Berwick-on-Tweed. Glacial action has created a series of raised beaches along much of the coastline.

Most of the hill land is made of shales and slates of Ordovician and Silurian age providing the distinctive smooth contours of much of the greatly folded Southern Uplands, including the isolated outcrops of the Moorfoots and the Lammermuirs, but immediately to the south of Edinburgh, the Pentland range is largely volcanic. The city itself is founded on a massive volcano, exemplified by the well-known rocks of Arthur's Seat in the middle of the Royal Park of Holyrood – the rock on which Edinburgh Castle sits is a volcanic plug, like that at North Berwick Law and elsewhere in this region. A broad band of carboniferous strata however stretches westwards from just north of Dunbar to cover much of the Central Belt, and provides the economic deposits of coal and limestones on which the mining and quarrying industry of the Lothians once depended.

While the hill land is moderately acid, the rich soils of East Lothian and Berwickshire provide some of the best farming land in the country where arable farming predominates in the lowland coastal belt. The same areas on the coast enjoy long periods of sunshine, enabling a wide variety of cereals to be grown, including wheat and high quality malting barley. Much of the previous sheep country of the hills has given way to extensive plantation forestry virtually to the skyline, but there is still extensive heather moorland used for grouse shooting in the low hills of the Lammermuirs and in the Tweedsmuir Hills north of Moffat. Because of their position and height, the latter are among the coldest of the hill masses in south Scotland and support a range of arctic-alpine species.

(Although just outside the region, the adjacent one Moffat Hills/Grey Mare's Tail site has been included here).

Acid upland moorland is also typical of the southern borders of West Lothian, which has also been subject to large-scale conifer forestry, and in contrast to the drier east, cattle and dairying are the dominant farming systems.

Population growth and industrial development has been concentrated around the commercial rock deposits in the north of the region, although the capital itself, Edinburgh, is not an industrial city. However, with the population very much concentrated here and in the immediate environs, agriculture in the vicinity has focused on the needs of the urban population, with development of vegetables, market gardening and soft fruits. The traditional woollen and weaving industries associated with the Border towns along the middle Tweed have declined considerably in recent years, and as elsewhere, deep mining has virtually disappeared. The relicts of the earlier oil shale mining can be seen in the pink-coloured bings of West Lothian, many of these now grassed over.

The good quality agricultural land encouraged innovation during the Agricultural Revolution in this region which produced a number of farming pioneers. The proceeds, often combined with revenues from mining and other industrial ventures, helped to finance the stately homes and fine landscaped estates which are a feature of the Lothians and Borders. Often, as in the case of **The Hirsel** near Coldstream and Hopetoun Estate on the Forth, and Dalkeith Estate in Midlothian, plantations of exotic trees were established and ponds created so that these mature estates are now rich in a variety of wildlife. Elsewhere, natural woodlands are few and far between, being confined to steep river gorges such as Roslin Glen outside Edinburgh or in the narrow ravines in the uplands where they are not available to sheep. In the hills, the reservoirs constructed to provide fresh water for the developing economic centres have become important havens for wildfowl, such as Gladhouse and Threipmuir.

One of the most important resources for wildlife is the coast, from the inner estuary of the Forth right down to the border. The Forth itself is of international importance for its wildfowl, notably its large populations of wintering duck, and its several islands, including the spectacular **Bass Rock** provide valuable seabird breeding sites, while dramatic coastal cliffs such as those found at **St Abb's Head** are renowned. The estuaries at **Aberlady Bay** and at the **John Muir County Park** near Dunbar not only provide safe roosts for wildfowl, but also support large populations of feeding waders along a coast which is remarkably unspoiled. Less well known are the series of wetlands south of Galashiels, from acid raised bogs to relatively rich fens, known as the Border Mosses, which are a feature of a landscape elsewhere poor in natural open waters other than the river systems such as the Tweed and its tributaries.

The historic and cultural attractions of Edinburgh need no repetition here, but it is also a city of many green spaces, not all of which have been manicured by civic authorities. Wooded parkland such as Corstorphine Hill, the scrubby margins of the Water of Leith flowing through the city and the crags and grasslands of Holyrood Park all support a variety of plant and birds including several relatively rare ones species. The great abbeys in the attractive border towns of Jedburgh, Kelso, and Melrose are not only among our finest ecclesiastical monuments, but are testimony to the riches of this fertile countryside as far back as mediaeval times, much of this based on the lucrative wool trade with Europe.

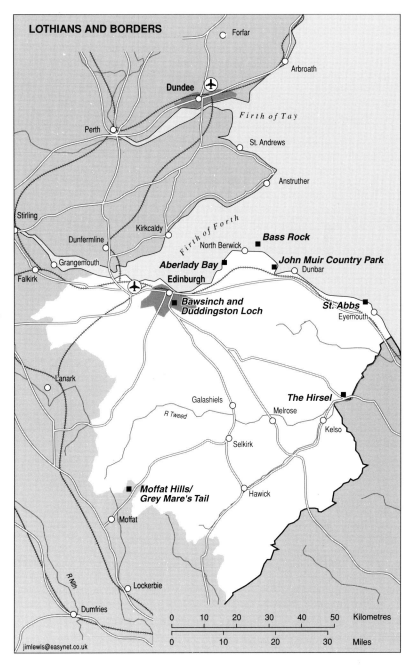

LOTHIANS AND BORDERS

Forfar

Arbroath

Dundee

Firth of Tay

Perth

St. Andrews

Anstruther

Stirling

Kirkcaldy

Firth of Forth

Dunfermline

North Berwick

Bass Rock

Grangemouth

Aberlady Bay

John Muir Country Park

Falkirk

Edinburgh

Dunbar

**Bawsinch and
Duddingston Loch**

St. Abbs

Eyemouth

Lanark

The Hirsel

Galashiels

Melrose

R Tweed

Kelso

Selkirk

**Moffat Hills/
Grey Mare's Tail**

Hawick

Moffat

R Nith

Lockerbie

Dumfries

| 0 | 10 | 20 | 30 | 40 | 50 | Kilometres |

| 0 | 10 | 20 | 30 | Miles |

jimlewis@easynet.co.uk

JOHN MUIR COUNTRY PARK *East Lothian*

This is one of the most diverse reserves in a county with a plenitude of fine coastal sites. The estuary of this river has developed long sandy spits where marshland surrounds rich mudflats, especially attractive to shelduck and such wintering waders as redshank, oystercatcher, ringed plover, dunlin, and knot in large numbers with smaller populations of turnstone and bar-tailed godwit. The sand spits provide good nesting cover for ring plover and for winter migrants. Both mute and whooper swans are present with a wintering flock of up to 80 birds, while wigeon, mallard and teal take advantage of good feeding around the marshland.

Other natural features of this country park are the steep sandstone cliffs at the eastern end, and the extensive sandy beach and dunes, backed by old plantations of pine and beech. There is an attractive walk through these woodlands to reach the shore at several points where areas of heather and cross-leaved heath have formed in the dunes. In summer, the wild thyme creates its own fragrance and the dunes support a variety of lichens. The beach is particularly good for shell collectors.

Off the A1, 4 miles west of Dunbar
L R
East Lothian Council

ABERLADY BAY East Lothian
East Lothian Council

The very first statutory nature reserve to be established by a Local Authority in Britain, this fine area of sand dunes and estuary has long been known as a bird watcher's paradise, of interest at any time of year. However, it is perhaps in the cold bright days of winter when it most comes into its own, with the tawny dune grasses offsetting the panorama of the Firth of Forth, and highlighting the undersides of the many wildfowl and waders which find rich feeding on the mudflats and surrounding marshlands. Among the 55 breeding bird species are eider, ringed plover and shelduck – the reserve is particularly noted for the last species.

The sight of over 3,000 pink-footed geese coming in to roost on a winter's evening after feeding on the fertile farmlands of East Lothian is an impressive one, but equally, the other wintering birds are of interest, especially the wigeon which can easily be seen in social groups feeding on the marshland close to the main coastal road. Waders are numerous and are also easily seen around the outflow of the Peffer Burn from the car park or roadside, including dunlin, grey plover, bar-tailed godwit, and knot whose silver undersides in flight create the effect of shimmering tinsel. There is much to interest the botanist, especially in the wetter areas between the sand dunes, with their orchids, grass of Parnassus, and many species of sedges.

Half mile east of Aberlady village on the A198
L R WC

BASS ROCK *East Lothian*

This is one of the great wildlife spectacles of the east coast of Scotland, a mere one hour drive from the capital city. All gannets take their Latin name *Sula bassana* from this site where they were first identified. The Bass Rock, an ancient volcanic plug rising sheer out of the Forth, is itself a dramatic sight even from the shore, where the wheeling gannets look like a swirl of snowflakes, but to appreciate its birdlife, the visitor needs to take the regular boat trips, running twice a day (weather permitting) from the attractive coastal town of North Berwick. With some 9,000 pairs of nesting gannets, this is a veritable Babel during the nesting season, when the close packed gannets are joined by kittiwakes, puffin, guillemot, razorbills, fulmars and shags. The sea arches are a good place to see either common or grey seals

From the harbour at North Berwick, check notice for sailings by Fred Marr.

BAWSINCH AND DUDDINGSTON LOCH *Edinburgh*

A surprising mini-wilderness less than 20 minutes from the heart of Edinburgh, this reserve lies on the south side of the Royal Park of Holyrood, dramatically situated below Salisbury Crag and Arthur's Seat, itself of international importance for its geological history, inextricably linked to the work of the great James Hutton. The loch itself is known especially for its pochard, which overwinter here, with up to 8,000 birds. Feral greylag geese also breed near the loch and the surrounding reedbeds provide a winter roost for yellowhammer and pied wagtail. Although the hide is available only to SWT members, there is open access to the north shore. At Bawsinch, the re-creation of a diverse woodland of native species and of small ponds has created a variety of habitats over a relatively small area, supporting nearly 70 species of birds and a rich woodland and wetland flora maintained by careful management.

Car park at eastern end of Duddingston Village.

L R *(SWT)*

Red Moss of Balerno *(SWT NT 164636)* Only 9 miles south of Edinburgh, this is the largest remaining raised bog in the district, with a variety of characteristic peatland species.

Jupiter Project *(SWT Near Grangemouth town centre, in Wood Street)* Scotland's largest wildlife garden has been created from a railway marshalling yard in Grangemouth, now landscaped with woodland, meadow and ponds. The site has a visitor centre and access to the reserve by wheelchair.

MOFFAT HILLS/GREY MARE'S TAIL *Borders*

In the rounded grassy hills not far from Moffat, a winding path leads upwards by the side of a steep ravine leading to Loch Skene, a high cold loch surrounded by rugged cliffs. Here one of the highest waterfalls in Britain falls some 60m alongside the winding path which takes the visitor to the heathery slopes above. The lime-rich cliffs around the loch are famed not only for their arctic-alpine flora, but also as the haunt of peregrine and raven hunting over this wild country. The sides of the gorge – which requires to be treated with great respect because of its steepness – has a lush vegetation especially where it is influenced by the spray from the waterfall. The walk is made colourful from late spring onwards by the purple of field scabious, the yellow of tormentil and the mauve of harebell. Here among the resident herd of wild goats, the 'billies' have characteristic long shaggy coats and impressive curving horns. Loch Skene resembles an arctic loch and has been chosen as one of the sites for the introduction of the arctic fish species, the vendace (see *The Vendace – An Ice Age Fish*).

On the A708, 15 miles north of Mofatt
L R *(NTS)*

ST ABB'S HEAD *Berwickshire (NTS/SWT)*

This reserve has some of the most spectacular scenery in south-east Scotland, with reefs, skerries and magnificent sea cliffs plunging sheer into the North Sea on this wild stretch of coast. It supports thousands of guillemots, razorbills, kittiwakes and puffins – up to 60,000 seabirds in the breeding months between April and late July. The birds can most easily be seen on the cliffs and stacks north west of St.Abb's lighthouse. There is a varied coastal plant life and offshore lies the only voluntary marine nature reserve in Scotland, protecting a beautiful underwater world in some of the clearest waters in Britain and one of the foremost diving locations in the country. At Petticoe Wick, there is an especially fine example of the folding of the Silurian strata which form the main rocks of the Borders.

Off the A 1107, 4 miles north-east of Coldingham
L R I WC

Pease Dean *(SWT NT 790705)* This densely wooded reserve of oak, ash, elm and sycamore, rises steeply above Pease Bay on the Berwickshire Coast, a rare surviving relict of previously extensive woodlands in this area. It is known for its localised insects and many non-flowering plants which thrive in the damp recesses of the ravine. There are paths throughout the reserve, including the east end of the Southern Upland Way.

THE HIRSEL *Borders*

In the intensively farmed lower Tweed Valley, among the richest wildlife habitats are the grounds and woodland policies of the larger estate houses, with their mixtures of tree and shrub species, often enhanced by artificial ponds and open parkland. This is an excellent example of a very old private estate, which, because of sympathetic management, has become a haven for a diversity of wildlife. It has been in the ownership of the Hume family for over three and a half centuries (and from 1166 was part of a Cistercian Monastery) and has not been subject to the dramatic changes which have affected the surrounding landscape. Its mixed woodland, parkland, farmland and open waters have no less than 160 bird species, of which almost 100 breed. On the Hirsel Lake, created in 1787 and the largest stretch of open water for 20 miles, up to 2,000 mallard roost, together with substantial numbers of shoveler duck, tufted duck and goldeneye. Other breeding bird species such as great spotted woodpecker, sedge warbler, stonechat, barn owl, dipper, and pied flycatcher reflect the range of habitats available. The original stable block has been converted into an attractive series of exhibitions on estate life and work and there are pleasant walks in the Leet Valley.

On A697 immediately west of Coldstream
L R WC I *

Duns Castle *(SWT NT 778550)* One of the great attractions of this SWT reserve is the spring flowers, with great sheets of bluebells and red campion, but it is probably equally well known for it variety of birdlife, notably for its small birds, including the quite local marsh tit, pied flycatcher (which nests regularly) and chiffchaff. Badger, roe deer and red squirrel are present in the mixed woodlands, and otters utilise the artificial loch, the Hen Poo, where yellow water lily, bulrush, and bog bean are prolific.

River Tweed *(NT 592628)* Between St. Boswells and Newton St. Boswells an attractive woodland path reveals a wealth of plants including marsh marigold, water avens, and butterbur by the riverside, and in the woodlands, meadow saxifrage, wood stitchwort, goldilocks, primrose and wild garlic.

Eyemouth/Coldingham *(NT 931660)* This fine 7 km coastal cliff walk in summer has splendid shows of sea pinks, sea bindweed and viper's bugloss, with a great variety of marine life in the shore pools, and offers an attractive addition to the reserve of St. Abb's to the north. Start either from Eyemouth car park or Coldingham Bay.

CHAPTER 21

South West Scotland

THIS REGION INCLUDES THE tourist areas of Dumfries and Galloway bordered by the Solway, and Ayrshire and Arran facing onto the Clyde, with the Clyde Valley including Greater Glasgow. The whole area is split by the Southern Upland Fault which separates the Ordovician and Silurian age shales and slates of the south from the Old Red Sandstone and Carboniferous sediment to the north, which are interrupted by volcanic rocks of various ages. In Dumfries and Galloway, occasional outcrops of granite such as at the Merrick and Criffel produce a more rugged highland scenery, but this is perhaps most dramatically displayed in the northern mountains of Arran – an island which, with elements of all of these strata, is a happy hunting ground for geologists. There is evidence of glacial activity everywhere, from the raised beaches round the coast to the drumlins of the Wigtownshire moors.

The main physical features are the Clyde Valley running right up onto the centre of this region, the hill country extending from east to west across Galloway, the low rolling moorlands extending from the Clyde Valley westwards and occupying much of Ayrshire, and a low-lying coastal strip belt of varying width around the border of the whole region. Glasgow itself lies in a basin surrounded by low hills. The climate is mild and damp, but the south-facing Solway Coast is noted for its sunshine record, although inland, the mountain country causes a marked rise in rainfall. Snow is rare throughout the region and rarely lies for any length of time. Although the soils of the Clyde Valley are relatively fertile, the poorly drained boulder clays of Ayrshire and elsewhere in the region often make for difficult farming conditions. Wigtownshire and other parts of Galloway are noted for extensive acid peaty bogland.

Land use is very much determined by these soil and climatic conditions, with stock rearing generally much more important than arable agriculture. On the better quality grasslands, both dairy and beef cattle are raised, and in the hills, sheep rearing predominates. However much of the ground previously dedicated to sheep has given way in recent years to some of the most extensive conifer forestry anywhere in the country, especially in Galloway. In the fertile and sheltered middle Clyde Valley, market gardening and soft-fruit growing are important, while in the lower Clyde Valley around Glasgow heavy industry predominated until World War II, but since then there has been considerable conversion to light industry. As elsewhere in Scotland, the coal and other commercial deposits which helped to make this area the "Workshop of the Empire" has declined, although open-cast mining is tending to replace deep mines. Salmon fishing is still a feature of the Solway and there is still some inshore and shell fishing from the small ports of Galloway and Ayrshire. Traditional tourism was based on the coastal resorts of Ayrshire, but Dumfries and Galloway, with its quiet charms and unspoiled pastoral countryside and coast, is becoming increasingly popular. The Solway coast also provides fine wildfowling, made even more popular by the improved rapid road access from the north of England.

With the possible exception of the huge flocks of wintering wildfowl on the Solway, this region cannot claim to provide comparable natural interest to that of Highland Scotland, with its spectacular mountain and loch scenery. Nevertheless it has many other attractions for the naturalist. The coast is an enormously varied one, from the great shifting sand and mudflats between the Rivers Lochar and Nith at **Caerlaverock** to the high cliffs at the **Mull of Galloway** and below **Culzean**. At Torrs Warren south of Stranraer, there are some of the highest and most extensive sand dunes in the whole country, while the shingle beach at **Ballantrae** is one of the finest examples of its kind. The warm south-facing Solway coast not only supports unusual coastal woodland and scrub, but provides good habitat for many southern species of plants and especially butterflies which are rare elsewhere in Scotland. Large areas of exposed intertidal coast have a wealth of wading birds. Freshwater wetlands are relatively scarce, but the **Ken-Dee** marshes in Galloway have outstanding bird life, rivalled only by **Lochwinnoch** in Renfrewshire with its record bird list.

While there are endless acres of softwood plantations, native woodland is scarce and usually confined to valley systems such as the Upper Nith, The Fleet with **Castramon Wood** and the **Cree** in the far south west, but there is a fine series of rich mixed deciduous woodlands along the banks of the middle Clyde at **Falls of Clyde** and **Nethan Gorge**. Elsewhere, old policy woodlands around great houses such as **Brodick** and **Culzean** reflect the achievement of the earlier aristocracy in beautifying their country estates. The **Galloway Forest Park** extends to 200 square miles, including both conifer and deciduous relicts, several of them important for their west coast Atlantic mosses and ferns in this damp climate. However the park also contains some fine hill country, including the granite massif of the **Merrick**, which along with other nearby summits such as the **Cairnsmore of Fleet** provides a habitat for predatory birds, including eagle, peregrine and raven, all of which have declined with the increase in commercial forestry and the loss of sheep carrion in this district. The hill country of both Galloway and **North Arran** provide sanctuary for wild goats in addition to red deer.

There is considerable historical interest in this region, including the very peaceful mediaeval abbeys at Dundrennan and Sweetheart in Galloway, and such spectacular examples of castles as Caerlaverock near Dumfries, and Bothwell on the Clyde near Hamilton, claimed to be the largest and finest stone castle in Scotland. The Wigtownshire town of Whithorn, the cradle of Christianity in Scotland, has again become a place of pilgrimage, while Glasgow Cathedral and the nearby St. Mungo's Museum of Religion, are now significant tourist attractions. Dumfries and several of the pleasant tidy rural vernacular villages of Galloway and Ayrshire have intimate associations with Robert Burns, and contrast with the more sophisticated cultural venues of Glasgow, with its Burrell Museum, Kelvingrove Museum and several art galleries of repute.

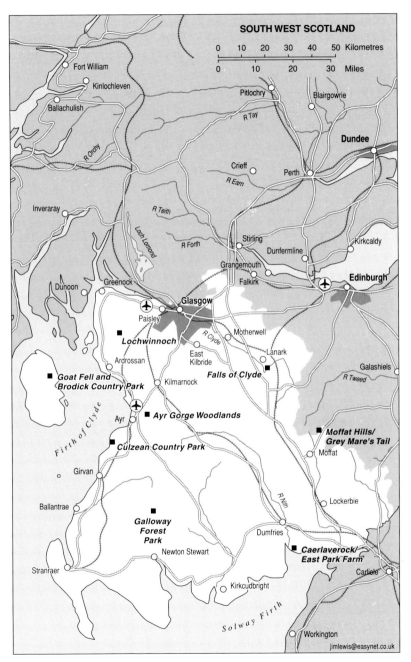

SOUTH WEST SCOTLAND

0   10   20   30   40   50 Kilometres
0   10   20   30   Miles

Fort William
Kinlochleven
Ballachulish
Pitlochry
Blairgowrie
R Tay
R Orchy
Dundee
Crieff
Perth
R Earn
Inveraray
R Teith
R Forth
Stirling
Kirkcaldy
Dunfermline
Loch Lomond
Grangemouth
Dunoon
Greenock
Falkirk
Edinburgh
Glasgow
Paisley
Lochwinnoch
R Clyde
Motherwell
Ardrossan
East Kilbride
Lanark
Galashiels
Goat Fell and Brodick Country Park
Kilmarnock
Falls of Clyde
R Tweed
Ayr Gorge Woodlands
Ayr
Moffat Hills/ Grey Mare's Tail
Culzean Country Park
Moffat
Firth of Clyde
Girvan
Ballantrae
Lockerbie
R Nith
Galloway Forest Park
Dumfries
Newton Stewart
Caerlaverock/ East Park Farm
Stranraer
Carlisle
Kirkcudbright
Solway Firth
Workington
jimlewis@easynet.co.uk

137

CAERLAVEROCK/EAST PARK FARM *Dumfries and Galloway*

By far the most interesting time to visit this reserve is between October and March when the great flocks of overwintering geese from the arctic are present, although Eastpark Farm (where there are special viewing facilities) holds a wide range of birds at all times of the year. The area is especially famed for its large populations of barnacle geese – the total breeding population from Spitzbergen – which arrive on the Solway each autumn, but there are also thousands of pink-footed geese and small numbers of greylag, as well as many duck and wading birds, notably pintail. There is a thriving colony of the rare natterjack toad breeding in the pools of the great saltmarsh, saved from extinction here by the artificial creation of ponds primarily for wildfowl (see *Natterjack Toad*).

The Wildfowl and Wetlands Trust have provided a series of viewing hides and towers for obtaining close up views of the huge flocks of birds which make this one of the most important international wintering areas. Caerlaverock has been recognised as a pioneer location for the harmonisation of controlled shooting, wildfowl conservation and traditional agriculture. On the margins of the reserve, the beautiful moated ruins of Caerlaverock Castle, with its warm red sandstone turrets and curtain walls, are an additional attraction.

Signposted off the B725, 12 miles south of Dumfries.

L R WC * I *(SNH/Wildfowl and Wetland Trust)*

Threave and Ken-Dee Marshes *(RSPB/NTS NX 7462 and NX 6376/6869)* Around the River Dee, the wetland and woodland site provides wintering ground for many wildfowl, including the largest flock of mainland Greenland whitefronted geese, and during the spring, migrant redstarts and pied flycatchers breed alongside the resident woodland birds. Greylag geese are common and bean geese and whooper swan are also seen, while pintail and shoveler nest regularly. Otters frequent this RSPB reserve and red squirrels are present. On the nearby NTS property, the marshes of Threave Wildfowl Refuge support feeding wildfowl, notably wigeon, mallard, teal and greylag geese, with a hide on the old railway providing a good viewing site. The 14th century Threave Castle, stronghold of the Black Douglas, adds an interesting historical dimension and a dramatic backdrop for hunting barn owl and wintering peregrine.

Mersehead *(RSPB NX 9255)* A new RSPB reserve covering a large area of wet meadows, saltmarsh, mudflats, and farmland, notable for its large flocks of barnacle geese, pink-footed goose, pintail and wigeon, with a good range of waders on the mudbanks.

GALLOWAY FOREST PARK *Dumfries and Galloway*

From the heights of the granite mass of the Merrick in the wonderfully named 'Range of the Awful Hand' with its eagles and ravens, to the RSPB oakwood reserve at the Wood of Cree, this park and its immediate surroundings have some of the wildest and most diverse scenery of this varied south-western region. There is a good chance of seeing red deer and especially wild goats, with their impressive curved horns and shaggy coats, in the sanctuary areas created for them, particularly in the enclosure on the A712 near Clatterinshaws, where guided walks commence from the Forest Wildlife Centre. At the highest levels, arctic-alpine plants such as dwarf juniper and starry saxifrage are relatively common, and mountain hare can be seen in winter, while lower down, the bogs are rich in sphagnum moss species, with plentiful bog myrtle and bog asphodel. The moorlands support hen harrier, and black cock on the woodland fringes, which the buzzards also prefer. Around Glen Trool there are surviving remnants of oak and birch woodlands, alive in summer with wood warbler, redstart, tree pipit and pied flycatcher. In the conifer woodlands, red squirrel and the re-introduced pine marten can occasionally be seen.

Off A714, 10 miles north-west of Newton Stewart
L R WC I *(FC)*

Mull of Galloway *(RSPB NX 157305)* On the most southerly point in Scotland, these impressive cliffs provide a habitat for many cliff nesting species, including cormorant, shag, kittiwake, razorbills and black guillemot. Gannets can be seen fishing offshore, and there are usually several puffins to be seen. The cliffs around the lighthouse are especially good for characteristic plants of this situation, such as spring squill and purple milk vetch, with a number of species here at their northern limit.

Castramon Wood *(SWT NX 592605)* A fine example of the old oak woods of the beautiful Fleet Valley north of Gatehouse of Fleet, with spectacular displays of bluebells in spring.

Cairnsmore of Fleet *(SNH NX 502671)* Open moorland on granite, important for upland animals and birds such as raven, with a large herd of feral goats. Peregrines continue to breed on this National Nature Reserve, but golden eagles are less successful. Although the grassland and moorland is relatively uniform, the unplanted hill from the valley of the Fleet to the heights of the Cairnsmore is an important oasis of open ground in an extensively afforested district.

AYR GEORGE WOODLANDS *Ayrshire*

An impressively steep ravine in the valley of the River Ayr, supporting ancient semi-natural woodlands, now one of the most important areas for wildlife in Ayrshire (according to local tradition, it was in these woods that Robert Burns plighted his troth to Highland Mary before they parted, never to meet again). Dominated by oak, it also has a tree and shrub cover of birch, hazel, holly, and rowan, while the acid ground conditions suit such plants as woodrush, bluebell and even heather and blaeberry. Many of the larger mammals inhabit the woodland, including, badger, roe deer and red squirrel, and the reserve is also noted for its uncommon invertebrates. Here the visitor may see great spotted woodpecker, kingfisher, dipper and grey wagtail. There is a good network of well-maintained footpaths and information boards.

At Failford, to the south of the A758 Ayr-Mauchline Road
L *(SWT)*

CULZEAN COUNTRY PARK *Ayrshire*

Surrounding Culzean Castle, the country park has impressive coastal cliffs of lava and old red sandstone, in addition to mixed policy woodlands and formal ponds. The woodlands provide habitat for roe deer, and red squirrel, and the bird life is extremely varied, including sparrowhawk, blackcap, and a diversity of warblers. The cliffs have many heathland plants such as tormentil and heather as well as typical coastal species – sea thrift and campion for example. The foreshore is a delight for children with its rock pools containing sea urchins, sea slugs, anenomes, brittle stars and butterfish, while agates are abundant. The coastal habitats support many of shore birds, and in winter, less common species such as long-tailed duck and great northern diver can sometimes be seen. Several pairs of shelduck use the sandy shore at Port Carrick. The park specialises in children's natural history programmes and has a variety of exhibitions.

Near Maybole, 12 miles south of Ayr on A719
L R WC * I *(NTS)*

Ballantrae *(SWT NX 0882)* A unusual shingle spit with a small colony of breeding little terns as well as common and arctic terns, ringed plover and oyster catcher. Meadow crane's bill and bird's foot trefoil to landward add to the shingle flora, which is also the site for the rare oyster plant.

GOATFELL AND BRODICK COUNTRY PARK *Arran, North Ayrshire*

Apart from its mountain wildlife, including the opportunity to see eagle, pere-grine, raven, golden plover and ptarmigan (at their most southerly station in Scotland) this massif has spectacular rock scenery. The highest peak on Arran, it is renowned for its geology and landforms, including massive sheer granite slabs and knife-edge ridges of jagged rock. The granite tors are the result of erosion of the original volcanoes and are an attraction for climbers and ridge walkers. On the higher altitudes are found such arctic-alpine plant species as alpine buckler fern, alpine lady's mantle, mountain sorrel, starry saxifrage, and dwarf willow, while lower down, the moorland supports heath spotted orchid, sundews and heath milkwort, and red deer, hen harriers and grouse are also frequently seen in this zone. Redpoll nest in the Glen Rosa woodlands. Brodick Country Park has old woodlands of both conifers and deciduous trees, with some especially fine examples of oak and beech. Wooded gorges and an attractive shore add to the range of wildlife, which includes red deer, badger, and offshore, seals and basking shark can be seen. The damp shady woodlands enable many mosses, liverworts, and lichens to flourish in this western location and to provide good habitat for chiffchaff , garden warbler, wood warbler and blackcap. Brodick Castle, with its formal gardens and fine interior collections, is well worth a visit.

2 miles (Brodick Country Park) and 5 miles (Goatfell) north of Brodick
L R WC I *(NTS)*

Glen Diomhan *(SNH NR 9246)* Although the rare native Arran Whitebeams are the justification for this National Nature Reserve, the walk up this typi-cal glaciated U-shaped glen may also be rewarded by sightings of eagle and peregrine.

Holy Island This small island off the east coast of Arran has white wild goats and Soay sheep, in addition to nesting ravens and buzzards on its cliffs. An ambitious programme of ecological restoration including tree-planting has been commenced by the Tibetan Buddhist community who use the island as a retreat centre.

Bute While most of this island in the Clyde is pastoral, the remoter north end, with its open moorland and rocky terrain, is known for its relict native wood-lands and a number of birds of prey, with the likelihood of seeing the resident herd of shaggy wild goats. Elsewhere on this island, the coast is noted for its scrub woodland, but large gatherings of duck and waders are present in the fine sandy bays in the south.

Ailsa Craig Although not usually accessible other than by organised excur-sion, this isolated rock in the Clyde west of Girvan is important primarily for its seabird colonies, notably for the large gannetry.

LOCHWINNOCH *Inverclyde*

Situated within the Clyde-Muirshiel Regional Park, this RSPB reserve is ideal for families, since all its facilities are readily accessible. It is one of the few remaining large open waters in this part of the west of Scotland and has interest throughout the year, with over 155 bird species recorded, 65 of which have bred. With its dense reed and sedge beds and wet woodland, there is a great variety of wetland habitat, providing for a diversity of birdlife. In winter, Aird Meadow and Barr Loch have abundant wildfowl, such as wigeon, pochard, mallard, teal, tufted duck and goldeneye, while greylag geese and whooper swans frequent the larger more open Barr Loch. In the spring, the fascinating mating display of the great crested grebes can be seen – there are around a dozen pairs of this attractive bird. The common species of waterfowl such as mallard, tufted duck, coot and moorhen with their broods can be viewed at close quarters from the hides, while sedge warblers are heard singing in the marshland, where yellow iris, marsh marigold, and marsh cinquefoil and several species of purple orchids add their own splashes of colour. Woodland plants such as wood sorrel, red campion, and ramsons occur in the fringing mixed woodland of oak, birch, beech, and lime. There are occasional sightings of otters. A programme of events for schools and the public is run throughout the year.

18 miles south-west of Glasgow between Paisley and Largs on the A760 near Lochwinnoch.

L R WC * I *(RSPB)*

FALLS OF CLYDE *South Lanarkshire*

This ancient gorge woodland surrounding spectacular waterfalls is best introduced through the SWT visitor centre at New Lanark, itself of considerable historical interest. The dramatic river landscape (readily seen via the attractive riverside walks) with its native oak, ash, and birch on the banks, supports such diverse species as kingfisher, grey wagtail, dipper on the river, great spotted woodpecker and willow tits in the mixed woodland, also used by roe deer, red squirrel, and badger, while kestrels nest on the steep side of the gorge. The woodlands have both common species such wood anemone, bluebell, dog's mercury, and a number of less common plants, including northern bedstraw and common wintergreen, while marsh marigold, cuckooflower and water avens grow by the river.

At New Lanark, 1 mile south of Lanark

L R WC I *(SWT)*

Moffat Hills/Grey Mare's Tail – see *Lothians and Borders*

CHAPTER 22

Central and West Scotland

THIS WIDELY DISPERSED REGION includes the tourist areas of Stirling, the Trossachs, Loch Lomond and Argyll with its islands. It is undoubtedly the most difficult region to describe coherently simply because of its sheer diversity - Argyll itself has probably the most varied landscapes in the whole of Scotland. The region covers everything from the largest single industrial complex in Scotland at Grangemouth on the inner Forth, with its vast petro-chemical complex, to the low windswept islands of Tiree and Coll, passing en route some of the most inaccessible mountain ranges on Jura, not to mention the undulating moorlands of **Islay** and the woodlands of the **Trossachs**. To the north, there are the impressive ranges of south Glencoe and the wilderness of Rannoch Moor, while the southernmost limits are, as the crow flies, 100 miles away at the Mull of Kintyre, with its tranquil pastoral scenery. There are both large and small islands, including Mull, Islay and Jura and smaller ones such as that natural gem, **Colonsay**. With such long sea lochs as Loch Fyne, Gareloch, Loch Long, Loch Striven and Loch Etive and the dramatic coasts of the islands, the seascapes are simply fantastic, while the region contains one of the longest freshwater lochs in Britain, in Loch Awe and Loch Lomond.

Split by the Highland Boundary Fault, the region offers prospects of both highland and lowland Scotland, the former mainly of the old Dalradian schists, slates and grits, while the central area is dominated by Carboniferous rocks, but also with some Old Red Sandstones in a band immediately to the south of the Fault. Some of the finest volcanic features in the whole of Europe are to be seen in this region, notably the outstanding example of tertiary volcanic activity on the central plateau of Mull, and the great lava flows which have produced the distinctive terraced appearance of much of the coast and the dramatic columns of Fingal's Cave on Staffa. The very obvious north-east to south-west bedding of the Dalradian rocks can perhaps be seen most clearly in the parallel outcrops across the width of Knapdale south of Lochgilphead. The strikingly green and fertile island of Lismore to the west of Oban is a good example of the influence of underlying limestone strata. Many of the coasts, but especially around the islands of Jura, Islay and Colonsay, have some of the finest examples of glacial raised beaches anywhere.

Dominated by the Atlantic and the warming influence of the Gulf Stream, the climate is generally mild and damp, with snow only lying for long in the north west mountains. Again the contrasts are remarkable –

the average annual rainfall at Inverary, at just over 2,000mm, is approximately double that on the island of Tiree, which claims some of the sunniest (albeit windy!) weather in the country. Mull, with its mountain mass facing on to the Atlantic, is known for its rainy climate. Most of the region is pastoral, with the farming dominated by sheep rearing, although cattle are also raised in the more fertile areas of the central districts and on Kintyre. With the exception of the parts of the Carse of Stirling and the Forth Valley, there is almost no high quality agricultural land suitable for arable cultivation.

These same conditions, notably the high rainfall, make much of this region eminently suitable for fast-growing conifer forestry, which covers large tracts of southern Argyll, as can be seen in the **Argyll** and **Queen Elizabeth Forest Parks**. However, considerable efforts are now being made in these areas and elsewhere in this highly scenic region to avoid the mistakes of the past, and to diversify these commercial plantations. It happens that the region also has the largest area of native woodland in the whole of Scotland, with fine examples within both of the Forest Parks, at **Glen Nant** east of Oban, and at **Inchcailloch** and **Inversnaid** on Loch Lomondside. Paradoxically, many of these rich native oak woodlands may have been preserved because of their economic importance to the charcoal industry in past centuries, with extensive coppicing and important centres based on Loch Fyne and Loch Etive.

Fish farming is the other resource industry which has grown dramatically in recent years, and is dependent on the clean sheltered waters of the long sea lochs, including both fin fish and shell fish, and which can be seen at **Taynish**. The margins of these sea lochs provide a remarkable diversity of ecological conditions from open sea coasts to almost freshwater systems at their landward extremities, attractive to animals such as seals, porpoises, and otters. At the other altitudinal extreme, arctic-alpine assemblages of plants can be found on the cliffs and ledges of the higher mountains such as Ben Lui and **Ben Lomond** where the resident wild goats are an additional attraction. The range of wetlands is demonstrated by the great expanse of raised bog at **Moine Mhor** bordering the Crinan Canal in Argyll and the famed **Loch Lomond**, recognised as a unique open water system, with its distinctly lowland and highland aspects. The islands are more renowned for their coastal attractions from the superb unspoilt beaches and sand dunes of Tiree and **Coll** to the wildfowl spectacles of **Loch Gruinart** on Islay.

The many other attractions of the region include Kilmartin Glen, near Lochgilphead, with which only Orkney can compare in its very accessible prehistoric monuments such as ritual stones and burial cairns, while the fort on the isolated eminence of Dunadd is reputedly the crowning place of the early Scots kings in this first kingdom of Dalriada. More well known is that beacon of Christianity of Scotland, Iona, off the southern tip of Mull and one of the most popular places of religious pilgrimage in the whole of Britain. In the east, Stirling with its fine castle and nearby Bannockburn provide a detailed insight into critical periods of Scottish history, including the Wars of Independence. Perhaps a surprising visitor attraction is the vast hydro-electric power station at Ben Cruachan at the head of Loch Awe, which is not only a most impressive engineering feat with its tunnels into the mountainside and huge turbine halls, but also a demonstration of the way in which ice-age corries have been utilised here and elsewhere in Scotland to provide 'power to the people'.

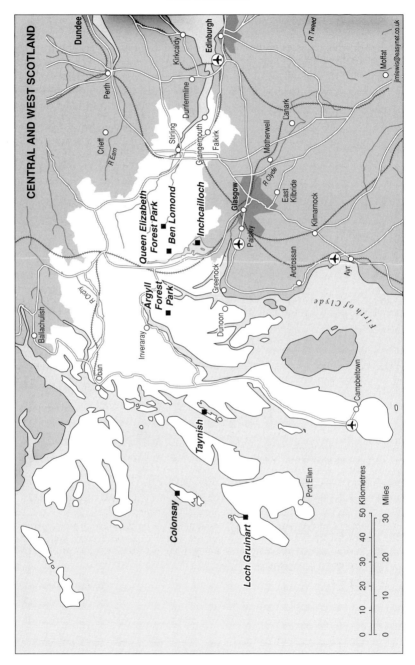

CENTRAL AND WEST SCOTLAND

Dundee

Edinburgh

Kirkcaldy

Moffat

jimlewis@easynet.co.uk

Perth

Dunfermline

Lanark

R Tweed

Crieff

Stirling

Motherwell

R Earn

Grangemouth

Falkirk

R Clyde

Queen Elizabeth
Forest Park

■ Ben Lomond

■ Inchcailloch

Glasgow

East
Kilbride

Kilmarnock

Paisley

Greenock

Ardrossan

Argyll
Forest
Park

Ayr

Dunoon

Firth of Clyde

Inveraray

Oban

Ballachulish

R Orchy

Campbeltown

Taynish

Colonsay

Port Ellen

Loch Gruinart

50 Kilometres

30 Miles

0   10   20   30   40   50

0   10   20   30

QUEEN ELIZABETH FOREST PARK *Stirling*

The park, encompasses mountain and moorland, forests, rivers and lochs. It is this combination which makes the area such an attraction for both wildlife and visitors, with a wide range of facilities to choose from, whether bird-watching, walking, mountaineering, riding or fishing. A particular feature is the Highland Boundary Fault, that great geological division of Scotland, which crosses the park and provides marked contrasts in the underlying rock types. Another is the old oak woodlands, formerly worked as coppice large-ly for the charcoal industry, and contrasting with the conifer plantations - the whole now being managed with recreation and amenity very much in mind. The park visitor centre *(NN 521015)*, off the dramatic Duke's Pass (A821) one mile north of Aberfoyle, is a good place to start, Here during May and June, nesting peregrines can be watched on closed-circuit TV.

From the visitor centre there is a range of well-marked and signposted trails of varying length illustrating the geology, forest use and wildlife of the area. Other longer walking routes take in several very beautiful small lochs such as Loch Drunkie, Loch Achray, and Loch Ard; one path provides fine views overlooking Loch Lomond, starting from the car park at Balmaha. Achray Forest provides a 7-mile forest drive with parking places and view-points. The drive runs alongside an area of wetland between Lochs Vennacher and Achray known as Blackwater Marshes where a good range of wildfowl are to be found, including greylag geese, goosander, teal, and wigeon. Given the size of the park area and its altitudinal range, it is not sur-prising that it supports a huge number of species, from those characteristic of the high tops, such as blue hare and ptarmigan, to old oak woodland at the lowest levels. With the removal of sheep these woodlands are now recovering their former glory and represent some of the most extensive native deciduous stands of trees in the whole of Scotland. They are noted for their populations of small birds, with pied flycatchers, wood warbler, redstart and tree pipit, as well as two species of woodpeckers.

The mountain areas have arctic-alpine plant species, such as the colour-ful moss campion and purple saxifrage and in the boggy areas, the insectivo-rous sundew and butterwort - often best seen along the wet ditches by forest roads, where ferns and mosses are usually abundant, and elsewhere there are extensive patches of the aromatic bog myrtle. On the drier areas, heather and other ericaceous plants dominate; these moorlands hold many skylarks, meadow pipits and curlew. On the lochs, pochard, goldeneye and tufted ducks are common, together with goosander and great crested grebe. Along the higher stream banks can be found the showy globe flower and starry sax-ifrage, and on slower moving stretches, grey wagtail, common sandpiper and dipper are usually present.

L R WC I * *(FC)*

BEN LOMOND *Stirling*

With its popular footpath starting from the shores of Loch Lomond at Rowardennan, the often steep (and muddy!) track up this famous mountain displays the diversity of habitats over this altitudinal range. Starting with the old deciduous woodland of oak around the loch shore, and quickly entering into rather uniform spruce plantations, the track leads to open grassy and heathery moorlands giving fine views across Loch Lomond. The open country which comprises the NTS property supports grouse and meadow pipits, while on the wetter peatlands, the sweet-smelling bog myrtle is characteristic. Just above the plantations an exclosure illustrates how native tree and shrub can recover from the previously intensive sheep grazing. As the most southerly Munro in Scotland, Ben Lomond has a good assemblage of arctic-alpine plants, especially on the ungrazed higher ledges. There is a well established herd of wild goats on the mountain.

From Rowardennan on the east shore of Loch Lomond, 10 miles north of Balmaha

L R WC *(FC/NTS)*

INCHCAILLOCH *Stirling*

Part of the National Nature Reserve which includes several of the other islands lying across the Highland Boundary Fault at the southern end of Loch Lomond, Inchcailloch can be reached by boat from nearby Balmaha. The nature trail leading to the wooded summit of the island provides the best way of seeing the varied habitats with ancient oak and mixed woodlands, including some fine old Scots pine. Within the woodland, there is the contrast in the rock types on either side of the faultline, so that lime-loving plants can be seen growing near to typical acid peatland mosses and ferns, for which the reserve is noted. Alder and ash grow in the damper areas, and in the richer pockets of soil, dog's mercury flourishes, alongside maidenhair spleenwort and sanicle. Small woodland birds such as willow, wood and garden warblers, are especially numerous, while redstart, great spotted woodpecker and jay also breed. From the summit of this island, it is possible in one direction to view the typical Highland scenery at the north end of Loch Lomond, and in the other, to see the great spread of the flatter agricultural lowlands of the Central Belt of Scotland.

L R WC *(SNH)*

Inversnaid *(RSPB NN337115)* On the east shore of Loch Lomond, this reserve has steep oak woodlands, renowned for their mosses and ferns – especially around the waterfalls, and for their populations of breeding flycatchers, wood warblers, tree pipits and redstarts.

TAYNISH *Argyll*

It would be difficult not to be entranced by the picturesque landscape and varied wildlife habitats of this wooded peninsula on the side of Loch Sween, well known for its rich marine life. Here is one of the largest oak woodlands in Britain, which has existed for the past 6,000 years, a mass of bluebells in spring, while its mild climate encourages a rich insect fauna, with over 20 species of butterfly – one of the best places in Scotland for both these and dragonflies. The woodland is especially known for its abundance and variety of non-flowering plants, typical of the high rainfall conditions of the west coast. Tree and shrub species include oak, ash, rowan, hazel and holly, with many of the tree trunks completely shrouded in lichens and mosses. Between the wooded ridges, wet meadows and boggy patches occur, with orchids such as heath spotted and northern marsh orchid. Buzzard and great spotted woodpecker are joined by seasonal arrivals such as redstart and wood warbler, which fill the woods with song in late spring. The shores of the loch provide a profusion of colourful seaweeds and other marine life, such as sea urchins, sponges, starfish – a favourite area for underwater divers in these warm sheltered waters around the peninsula. Both otter and common seal are present in Linne Mhuirich.

12 miles west of Lochgilphead off the B8025
L R *(SNH)*

Moine Mhor *(SNH NR 8292)* The 'Great Moss' near Lochgilphead, is best seen from the height of the ancient hill fort of Dunadd where its extent and unusual situation adjacent to the Crinan Canal can be appreciated. Now a National Nature Reserve, it has explanatory information boards describing the specialised plants and insects, such as two species of cranberry and the nationally rare large heath butterfly. Hen harriers and ospreys are seen hunting from time to time. A board walk from the west side leads on to the moss.

Fairy Isles *(SWT NR 766884)* Six small wooded islands and a narrow strip of coastal broadleaved woodland, at the head of Loch Sween, which attract a variety of coastal birds, including terns, eider, heron, and red-breasted merganser.

Carradale *(SWT NR 8137)* A good place to look out for a variety of marine life, including otter, white-beaked dolphin, bottle nose dolphin, porpoise and killer whale, all of which are regularly seen. Both sika deer and wild goats are present, and many kinds of seabirds.

LOCH GRUINART *Islay*

It is possible to view the great flocks of wintering barnacle and Greenland white-fronted geese from the comfort of your car on the B807 which straddles this reserve at the north end of Islay. However at the RSPB visitor centre, a live video camera link lets you get even closer views of the 25,000 grazing geese, which share the area with hen harriers, buzzards, short-eared owls, peregrines, merlin, and golden eagles. The farmland and moorland of this reserve around the loch provide one of the two most important wintering areas for barnacle geese in Britain (the other being Caerlaverock on the Solway) with approximately 20,000 birds making up two-thirds of the total Greenland race, while the Greenland whitefronts represent a third of the world population of this species. The farmland, saltmarsh and sandflats also provide a sanctuary for waders and duck, especially teal, shelduck, and eider. During the spring, there are hundreds of breeding waders – lapwing, redshank, and snipe, and the inimitable call of the corncrake is heard in the evenings. On the moorland, curlew and both black and red grouse breed.

Islay has every sort of habitat available on the west coast of Scotland, from loch to hilltop, from sea cliff to marsh. Choughs are a feature of this coast especially at **Mull of Oa** and both the goose species above also congregate at **Loch Indaal**, which has a huge wintering flock of eider duck. Altogether, over 225 species of birds have been recorded on the island, of which 110 breed.

Loch Gruinart reserve Signposted from the A8477 Bridgend to Bruichladdich road, 3 miles from turn-off. (Islay is reached by ferry from Kennacraig to Port Akcaig or Port Ellen or by air from Glasgow)
L WC R I* *(RSPB)*

Coll *(RSPB NM1554)* The typical Hebridean coastal habits can be found here – long white-shell beaches, high sand dunes and flowery machair grassland. This RSPB reserve is managed primarily for it corncrakes which have been declining generally elsewhere, but there many other breeding birds, including redshank, lapwing and snipe. Large number of barnacle and Greenland white-fronted geese come to the island.

The Burg *(NTS NM 426266)* On the north shore of Loch Scrideain on Mull, this site involves a hard 7-mile coastal walk to McCulloch's tree, a giant fossil estimated to be at least 50 million years old. En route there is a wealth of natural history interest including a profusion of butterflies attracted by the rich grassland flora.

COLONSAY *Argyll*

Although each of the Inner Isles is unique in its own way, the island of Colonsay is especially varied in its wildlife habitats, from remote and rugged moorland at its northern end to thrift-covered saltmarshes in the south. The ancient Torridonian sandstones have been sculpted into remarkable shapes and there are fine examples of raised beaches. The shallow lochs and their fringes are bright with bogbean and waving cottongrass and otters are not uncommon. The island supports some of the oldest native woodland – rarely exceeding 6m in height – in the west of Scotland, mainly of oak, ash, rowan, and birch: underneath, in the carpet of moss, there is a rich variety of flowering plants and fungi – more than 620 native plants have been recorded from the island, in addition to 550 fungi and a staggering 300 lichens, while a similar number of moths and butterflies have been recorded. This is one of the easiest places to see wild goats, which frequent the coastal grasslands and moorlands. About 90 species of birds breed: over the last 100 years, some 200 species have been seen, including some rare migrants. With its spectacularly beautiful beaches and dunes, as well as dramatic seabird cliffs, it would not be an exaggeration to describe the island as a naturalist's paradise, but there is also considerable historical interest from Neolithic times onwards, including the association with St. Columba and the remains of a 14th century Augustinian Priory. Illustrated talks on the island's natural history are provided regularly throughout the summer at the Colonsay Hotel.

Car ferry from Oban 3 times each week.

Glen Nant *(FC/SNH NN 0128)* This native deciduous woodland has long been known for its association with the charcoal industry of the district, especially with the nearby Bonawe Furnace where Historic Scotland provide an excellent exhibition on this important local industry. Here the coppice oaks used for this purpose are very evident, as well as the charcoal hearths which can still be seen. Apart from the oak which was favoured for this purpose, there is now birch, ash, hazel, bird cherry, rowan and holly, wych elm, and in the wetter areas, alder and willow. The reserve holds good breeding populations of typical birds such as wood warbler, redstart, and great spotted woodpecker. Under these heavy rainfall conditions, ferns and mosses are abundant, often carpeting the ground and fallen tree trunks alike.

Inverliever *(FC NM 9410)* A mixture of open moorland, oakwood and plantation forest on the side of Loch Awe supports a number of predatory birds, including nesting hen harrier on the open moor, and in the forest, breeding sparrowhawk, buzzard, alongside jay and crossbill. Red, roe, and sika deer and pine marten and badger are present.

ARGYLL FOREST PARK *Argyll*

Lying entirely north of the Highland Boundary Fault, this forest park is the only one penetrated by deep sea lochs – Loch Long, Loch Goil and Loch Eck reaching out into the Firth of Clyde, and forming several great peninsulas of forest and hill, often dropping dramatically down to the coast. About half of the park is planted, mainly with conifers, many of which have grown to considerable size, while those in the arboretum at Benmore are gigantic specimens, flourishing in one of the best climates in the whole of Europe for such North American Pacific species.

There are well over a score of hills above 600m, the highest reaching 926m on Beinn Narnain at the head of Loch Long. These are formed of ancient Highland shales, sandstones and mica-schists sandstones, often massively folded and contorted to create the well known pinnacles of the Cobbler above Arrochar. Much of the park receives over 2,500mm of rain so that the rocks are frequently dripping with water and treacherous underfoot. The same rain creates a multitude of waterfalls, both large and small, and which combined with the mild temperatures, favours the exotic rhododendrons and azaleas – more than 250 of them – of the Benmore Gardens. Along the shore of the sea lochs there are great flocks of wading birds: curlew, oystercatcher, sandpiper, redshank, turnstone and ringed plover. Further out, flotillas of eider are common, alongside red-breasted mergansers and goosanders, which nest around the forest lochs. Porpoise and otter are often seen.

The mixed broadleaved woodlands are mainly found above the loch shores and in the many secret valleys – oak, ash, birch and hazel. Primrose and violet are often the earlier spring flowers, followed by great sheets of bluebells in the open glades, with wood anemone and sorrel. The birds of these woods include redstarts, tree pipits, pied flycatchers, wood warblers and jays. By contrast, the conifer plantations in their early years favour buzzard, kestrel, hen harrier, and short-eared owl as a result of the upsurge in voles following fencing and exclusion of sheep. In the more mature plantations, goldcrests and coal tits feed, and also those cone seed eaters, the crossbill and siskin, while great spotted woodpecker and tree creeper are attracted by dead and dying timber. The forest cover provides sanctuary for wildcat, fox and red squirrel.

Above the treeline, the drier grasslands and moorlands are interspersed with wetter areas, often dotted with orchids in spring and early summer, and the bog asphodel, which in autumn turns these areas orange. Both the sundew and butterwort find plentiful insects here to sustain them, not least the ubiquitous midge of these western areas. All the mountain birds are here – ptarmigan, golden eagle, raven, hoodie crow, red grouse and peregrine.

Accessible at various points from A815, B839, B 828 and A83
L R WC I *(FC)*

CRINAN WOODLANDS

A fine old deciduous woodland of oak, ash, birch and hazel in a very scenic location overlooking the Crinan Canal. A delightful woodland walk leads to a viewpoint with spectacular views across to Jura and other inner Hebridean islands. Buzzards, tree creepers, wood warblers and redstarts are all present, as well as roe deer and red squirrels, while butterflies such as the speckled brown and Scotch argus are frequently seen. The surrounding area is renowned for its historic interests, notably prehistoric burial centres and stone circles.

WT *(NR 790938)*

THE WEST HIGHLAND WAY

This was the first long-distance footpath to be established in Scotland (the others are the Southern Upland Way and the Speyside Way) and runs for 152 km from Milngavie immediately to the north of Glasgow to Fort William –
    "from Scotland's largest city to the foot of its highest mountain, along the shores of its grandest loch across its grandest moor"
– according to the official guide. First conceived more than 50 years ago, it is now a fine walk through some of the most varied and splendid scenery of Scotland, in parts a strenuous and rough trek with considerable risk of exposure if the weather turns nasty. There are sharp contrasts in geology and landscape as the walker passes from lowlands to highlands, as well as in land use, human activity and wildlife. Much of the route follows old droving roads and historic highways, including a number of the military roads and bridges built to police the Highlands in Jacobite times, as well as old and now disused railway lines. Passing as it does through woodland, along loch shores, across high moorland and through spectacular rugged mountain country, the walker is likely to encounter the wide range of wildlife which these contrasting habitats support, not least in summer, the notorious highland midge!

NS 556744 to NN 113743

CHAPTER 23

National Scenic Areas

THESE ARE THE AREAS which because of their outstanding landscapes and physical features have been designated as National Scenic Areas, and most lie within the Highland region. As indicated in the introductory section on Protected Land in Scotland, 40 areas have been officially designated for their outstanding landscapes (see map). In almost all of these areas there are other sites of wildlife interest, many of them of the highest quality and which are briefly described in the foregoing section on *Places to Enjoy Wildlife*. Usually much larger that the sites indicated above, the NSAs frequently contain such smaller designated areas, and may have within their boundaries several National Nature Reserves and other conservation areas. They range from small examples of fine Southern Upland scenery such as the distinctive Eildon Hills near Melrose to some of the wildest and most remote land in Britain in the so-called Rough Bounds of Knoydart. The areas of the north-west in particular offer the nearest approach to a "wilderness experience" to be found in this country for those with the fitness and competence to to penetrate on foot beyond the localities which have public access and information facilities.

Many national scenic areas do not have marked boundaries and it is assumed that visitors who wish to enter these areas beyond the public roads system will be competent in the use of map and compass. Apart from the general location map therefore, no other access reference is given. However, a number of these areas, especially in the south, have public roads within and around their boundaries, so that more modest expeditions can be made using the facilities provided by various conservation agencies, and they can also provide visual delights even by car-borne travellers, for example the very fine drive along Loch Maree through the very heart of the National Scenic Area of Wester Ross, while even the heights of the Cairngorms are readily accessible by ski road and chairlift.

Because of their importance as examples of the finest landscapes in Scotland and their inclusion of a number of other important wildlife sites, a selection of the designated NSAs (particularly areas outside the north-west which have not been described previously) are listed here with basic details of their location, main features, and accessibility. The references in italics are to specific areas previously described under *Places to Enjoy Wildlife*.

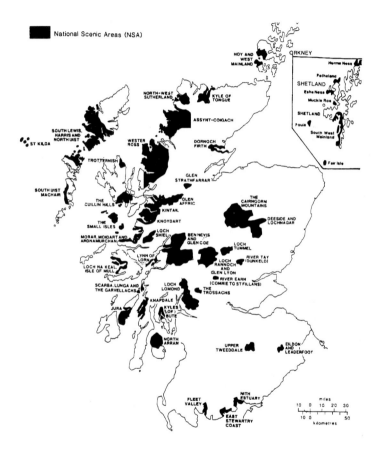

National Scenic Areas (NSA)

ORKNEY

HOY AND WEST MAINLAND

Herma Ness

Fethaland

SHETLAND

Esha Ness

Muckle Roe

SHETLAND

Foula

South West Mainland

Fair Isle

NORTH-WEST SUTHERLAND

KYLE OF TONGUE

ASSYNT-COIGACH

SOUTH LEWIS, HARRIS AND NORTH UIST

ST KILDA

WESTER ROSS

DORNOCH FIRTH

TROTTERNISH

GLEN STRATHFARRAR

SOUTH UIST MACHAIR

THE CUILLIN HILLS

GLEN AFFRIC

THE CAIRNGORM MOUNTAINS

KINTAIL

DEESIDE AND LOCHNAGAR

THE SMALL ISLES

KNOYDART

LOCH SHIEL

MORAR, MOIDART AND ARDNAMURCHAN

BEN NEVIS AND GLEN COE

LOCH TUMMEL

LYNN OF LORN

LOCH RANNOCH AND GLEN LYON

RIVER TAY (DUNKELD)

LOCH NA KEAL, ISLE OF MULL

RIVER EARN (COMRIE TO ST FILLANS)

SCARBA, LUNGA AND THE GARVELLACHS

LOCH LOMOND

THE TROSSACHS

JURA

KNAPDALE

KYLES (OF) BUTE

NORTH ARRAN

UPPER TWEEDDALE

EILDON AND LEADERFOOT

FLEET VALLEY

NITH ESTUARY

EAST STEWARTRY COAST

miles

10 0 10 20 30

10 0 50

kilometres

Kyle of Tongue

Around the sea loch of the Kyle of Tongue, this area in north-west Sutherland is dominated by the isolated peaks of Ben Hope and Ben Loyal forming a stunning panorama of peaks on the skyline at the landward head of the loch. The varied woods and crofting settlements add variety to the coastal scenery, which is exemplified by the great sweep of sand and dunes at the mouth of the River Naver. A minor road runs round the head of the loch, but otherwise, access is across rough moorland and peat bog *(see Invernaver)*.

Nearest town: *Tongue, Bettyhill*

Assynt-Coigach

Some of the most rugged and unusual scenery in Britain, with seven well-known mountains, including Stac Pollaidh, Suilven, and Benmore Coigach among others providing an unparalleled upland landscape. The mountains are known for their fantastic shapes rising out of a rolling rocky landscape of ancient Lewissian gneiss, with ridges of white quartzite contrasting with weathered red sandstone. The seaboard is equally dramatic, with its pattern of offshore islands, inlets, sandy bays and crofting settlements. Access to the interior is difficult, but there are marked paths provided at the reserves mentioned here *(see Ben More Coigach, Inchnadamph and Inverpolly)*.

Nearest Town: *Ullapool*

Wester Ross

Sometimes described as the last great wilderness of Scotland, this vast area, the largest of all the NSAs, has both rugged mountain scenery and much gentler landscapes round the lochs and coast. The great jagged massif of An Teallach, a favourite of climbers, dominates the north eastern sector but is paralleled by the huge soaring mound of Liathach in Glen Torridon in the south. Loch Maree, with its islands and surrounding pinewoods, is renowned for its beauty, while Loch Torridon sets off the forbidding grandeur of the Torridon Mountain mass in the south. The area is studded with attractive lochs, often with woods of birch, pine and oak on their shores, while the coastal scenery is as diverse as anywhere in the west, including the famed Gruinard Bay. Access is best via the reserve trails at Beinn Eighe and elsewhere *(see Beinn Eighe, Torridon, Corieshalloch Gorge, Allt nan Carnan)*.

Nearest towns: *Kyle of Lochalsh, Ullapool*

Morar, Moidart and Ardnamurchan

This is a coastal landscape like no other, around a series of sea lochs such as Loch Sunart, Loch Moidart and Loch Ailort. Although framed by rugged mountain scenery, this is a very intimate coastal landscape of great complexity, with its inlets and small islands, many of these wooded, with spectacular views across to the Small Isles. In the north, around the Sound of Arisaig the coast is much indented with rocky shores and sandy bays on either side of Loch nan Uamh, while in the south another prospect is provided by the flat sandy bay and great peat moss of Kentra. Although the hill country is for the fit, there is fine coastal walking around the many bays and inlets from the main coast road.

Nearest town: *Fort William*

Knoydart

Probably the most remote of all the NSAs, Knoydart is circled by quite forbidding mountain scenery enclosing at north and south ends respectively, Lochs Hourn and Loch Nevis, which penetrate deep into this wild country. The first of these lochs has been described by one highland writer as

"the lake of the infernal regions – a wild country of dark sea lochs and gloomy hills, often mist shrouded".

Ladhar Beinn (owned by the John Muir Trust) with its very toothed alpine outline is at 1,020m the most westerly Munro in Scotland, its clearly pointed summit rising above the great hollow of Coire Dhorccal, lying in a horseshoe of ridges and buttresses, from where there are fine views across to Skye. The area includes some scenic coastland along the shores of the Sound

---

Did you know that:

* Open air recreation generates £730 million pounds for the economy ?
* This activity represents 29,000 full-time equivalent jobs ?
* There are 7,850 Scottish jobs directly related to its natural heritage ?
* Walking in the countryside is the most popular leisure activity ?
* About 50,000 people annually complete a long-distance route in Scotland ?
* In 1997, some 94,000 people took a walk along Long Distance Footpaths in Scotland ?
* In the same year, an estimated 23 million walks were taken in the Scottish countryside ?
* The next long-distance route will link Inverness and Fort William along the Great Glen ?
* Due to public use more than £700,000 is required to repair the West Highland Way

of Sleat. Despite its apparent isolation, there is plenty of evidence of human habitation in the ruined crofts and fishermen's cottages. Access to this remote area, which has no internal roads, is either by ferry from Mallaig or at the eastern end by road starting at Kinloch Hourn.

Nearest town: *Mallaig*

South Lewis, Harris and North Uist

This is a very large area embracing the highest peaks in the Outer Hebrides in North Harris, and the great sweeps of sandy machair along the North Uist shores, while in-between there is much rolling moorland, bog and lochan. The hills are as rugged as any on the mainland, often with precipitous crags and much exposed ancient Lewissian rock. The east coast of Harris is quite unique in its deeply dissected knock-and-lochan topography, with innumerable bays and islets, supporting tiny crofting communities. By contrast, the wide sandy beaches of the west coast, often contained within rocky promontories dropping steeply to the sea, stand out against the dark peaty inland moors. Long sea lochs mean that the presence of the sea is always close at hand. This is an almost entirely treeless landscape, whose very barrenness is part of its attraction, and where the elements of sea, freshwater, rock and moor are intimately woven together. Much of the interior, especially in the Harris hills is hard walking over rough country, but most of the coast is readily accessible from the road system *(see Balranald, Loch Druidibeg)*.

Nearest town: *Tarbert*

River Tay (Dunkeld)

One of the smallest NSAs, the area comprises the area in the immediate vicinity of the delightful old Royal Burgh of Dunkeld, with its ancient cathedral and attractive historic houses. Lying on the Highland Boundary Fault, from the south the sudden ruggedness of the surrounding hills comes as something of a surprise. These hills have a pleasing variety of native and introduced trees, mainly coniferous, often clinging to steep rock faces. Much of the attraction of the area lies in the sweep of the River Tay in and around Dunkeld and the Braan, with the contrast between a wide, smooth- flowing river, and the rapids and waterfalls elsewhere, with many small lochans nearby. In a small area, there is a great diversity of scenery, with wildlife and historic interest set in the popular tourist district of central Perthshire district. There is a whole network of maintained paths and marked walkways in the area *(see Loch of the Lowes, Tummel Shingle Islands)*.

Nearest town: *Dunkeld, Perth*

Upper Tweeddale

Immediately east of Peebles, this small NSA takes in the upper course of the River Tweed where it is contained within a narrow steep-sided valley of considerable scenic attraction. Much of this is due to the contrast between rich well-kept riverside farmland, varied deciduous and conifer woodlands and upland pasture, not to mention the winding river itself. Along the valley, there are ancient castles and mansions, tidy farms, and historic churches, all combining to give this prosperous landscape a long-settled look. The old burgh of Peebles with its wide main street and pends leading off represents a typical established Border town on the banks of this famous river, which provides attractive riverside walks, while the surrounding forests and heather moorland offer good hiking opportunities.

Nearest town: *Peebles, Edinburgh*

Wildlife Viewing Services

THE FOLLOWING IS A SELECTION of commercial services, excluding nature reserves, where the purpose is primarily wildlife viewing – there are many other suppliers such as boat trips, where this is incidental. As there are many more operators than can be included here, a selection has been made to provide the widest geographical distribution. The description is abstracted from informa-

WILDLIFE GUIDES

WILD COUNTRY EXPEDITIONS
David Hassan 01389 731875

Exploration of some of Scotland's remoter islands.

ISLE OF MULL WILDLIFE EXPE-DITIONS
Ulva House Hotel, Tobermory, Isle of Mull PA75 6PR 01688 302044

Guided tours to view wildlife, such as otters, golden eagles and sea eagles, which visitors would only expect otherwise to see on TV. Habitat protection is encouraged.

LEWIS COUNTRYSIDE SERVICES
20 Hacklete, Great Bernera, Isle of Lewis HS2 9ND 01851 612288

Guided walks and tours enabling clients to get more out of their visit by providing information on wild-life, history and culture.

SPEYSIDE WILDLIFE
Fearn Dell Road, Nethybridge, Inverness-shire PH25 3DG 01479 821478

Organises week-long tours for wildlife enthusiasts. Visitors see and experience much more wildlife and landscape than if they had been on their own.

GLENLYON WILDLIFE SAFARIS
Donald Riddell, Glen Lyon, Perthshire 01887 877235

Land Rover safaris take place on a full-day, half-day or evening basis offering an opportunity to view Highland wildlife in its natural habitat.

ISLAND ENCOUNTER WILDLIFE SAFARIS
Gruline Home Farm, Aros, Isle of Mull PA71 6HR 01680-300437

Uses an 8-seat vehicle fitted with binoculars and wildlife books and accompanied by an expert guide to see the wilder and more beautiful places on Mull and its abundant wildlife.

BIRDWATCHING BREAKS
26 School Lane, Herne CT6 7AL 01227 740799

Specialised holidays in finding birds, their habitat and environs, visiting reserves, SSSIs and other important areas.

HEATHERLEA BIRDWATCHING SERVICES
Deshar Road, Boat of Garten, PH24 3BN 01479 831674

Birdwatching holidays based in Boat of Garten. Expert guide will lead you to the habitats of all the key Scottish species.

SHETLAND WILDLIFE TOURS
Fairview, Scatness, Virkie, Shetland ZE3 9JW 01950 460254

Offers a wide range of tours throughout the year to show beginners and experts a range of Shetland's wildlife.

HIGHLAND SAFARIS
Cnocmor, Strathpeffer IV14 9BT 01997 421618

Provide walking holidays lasting one week to ten days in small groups, journeying throughout the Highlands and Islands

GREAT GLEN WILDLIFE
Sherren, Harray, Orkney KW17 2JU 01856 761604

Tours are tailored to provide opportunities to experience the maximum number of species that may be seen in a limited period. Trips also include discussion on land use - so that clients have a better understanding of practices.

MERLIN WILDLIFE TOURS
Sonnhalde, Kingussie, Inverness-shire

Informal discovery of the wildlife habitats of the Scottish Highlands. Also provide photographic weeks.

JOHN CARRUTHERS
Ness Farmhouse, Ness Road East, Fortrose, Ross-shire IV10 8SE 01831 621545

Provides a relaxed, easy-pace holiday viewing the wildlife visitors have come to see.

ANDREW CURRIE, NATURALIST
Glaseilean, Harrapool, Broadford, Isle of Skye IV49 9AQ 01471 822344

This naturalist/writer leads a number of parties each year on organised walks to introduce them to the natural history and scenery of the area.

GUIDED LANDROVER SAFARIS, JURA
The Isle of Jura Hotel, Craighouse, Jura 01496 820243

5 to 6 hours tour, including a 2/3 mile trek to see the famous Corryvreckan Whirlpool. Trips by appointment only.

MORAY FIRTH DOLPHIN AND SEAL CENTRE
North Kessock, Inverness. IV1 1XB 01463 731866

A great place to watch dolphins in the Kessock Channel just minutes from the centre of Inverness. Listen to Dolphin noises through the unique hydrophone system.

HOLIDAYS WITH WILDLIFE

THE HIGHLAND FIELD CENTRE
Strathconon, Ross-shire IV6 7QQ 099 77260

Environmental education centre offering activity programmes for schools, unaccompanied children and organised groups of all ages.

ORKNEY ISLAND WILDLIFE
Shapinsay, Orkney KW17 2DY 01856 711373

Orkney Island Wildlife provides an excellent holiday leading to increased awareness of environmental issues and the range of wildlife in the Orkney archipelago.

KINDROGAN FIELD CENTRE

Enochdhu, Blairgowrie, Perthshire
PH10 7PG 01250 881286

The Centre provides field study courses to various locations, mainly in the Highlands and Islands of Scotland. The aim is to increase the knowledge and understanding, awareness and appreciation of all participants.

VIEWBANK GUEST HOUSE

Golf Course Road, Whiting Bay, Arran 01770 700326

John Reid, author of 'Birds of Arran', offers birdwatching holidays based on Viewbank Guest House.

ISLAND HOLIDAYS

Drummond Street, Comrie, Perthshire PH6 2DS 01764 670107

Provide seven night guided tours to, for example, Shetland, to view the abundance of wildlife in these islands.

AIGAS HOUSE & FIELD CENTRE

Beauly IV4 7AD 01463 782443

The Field Centre provides organised holidays and study tours with a wildlife/cultural focus. These are led by expert tutors.

STROME HOUSE

North ,Wester Ross 01520 722588

Mike Scott is a botanical expert and takes groups of visitors to see the flora and fauna of the local area.

BOAT TRIPS

SEA SAFARIS

St Marys, Tobermory, Isle of Mull PA75 6PN 01688 302111

MV Kelowna, provides visitors with opportunities to view whales and visit remote islands.

SEA LIFE SURVEYS

Dervaig, Isle of Mull PA75 6QL 01688 400223

Sea Life Surveys offers a variety of trips to view the whales and dolphins off the west coast of Scotland. The crew are all wildlife specialists. The MV Alpha Beta is a 40ft trawler yacht. The firm also has lodge accommodation.

STAFFA TRIPS

Tigh-na-Traigh, Isle of Iona, Argyll PA76 6SJ 01681 700358

Boat trips (April to October) to view Fingal's Cave and puffins on Staffa. Occasional sightings of cetaceans on way. Other species pointed out by skipper using identification card.

ARDNAMURCHAN CHARTERS

Laga Bay, Acharacle, Argyll PH36 4JW 01972 500208

Sea trips including opportunities to view the area's abundant wildlife.

MV VOLTAIR

Lagg Cottage, Dervaig, Isle of Mull PA75 6QY 01688 400380

The Eastwoods take visitors for a week afloat, seeing wildlife in its natural habitat.

TURUS MARA

Penmore, by Dervaig, Isle of Mull PA75 6QS 01688 400242

Provides a variety of whole or part day landing cruises by experienced guides/skippers who share their knowledge and some lesser known facts about the prodigious island wildlife.

LORNE LEADER LIMITED

Ardfern, Lochgilphead PA31 8QN 01852-500212

Six day wildlife cruises around the Hebrides on a large traditional sailing boat, led by an expert naturalist with an intimate knowledge of the area.

WILDLIFE CRUISES ON LOCH ETIVE
01866 822430

Cruises to Loch Etive's secret places to see a wide variety of wildlife.

BRUCE WATT SEA CRUISES
The Pier, Mallaig, Inverness-shire PH41 4QS 01687 462320

Boat trips to view wildlife of area.

DOUNE MARINE
By Mallaig PH41 4PU 01687 462667

Operates three boats taking visitors to view wildlife. It also provides organised trips to islands such as Rhum and St Kilda, by agreement with SNH and the National Trust for Scotland.

MINCH CHARTERS
Harbour Slipways, Mallaig 01687 462304

The MV Cuma carries 12 passengers in six twin cabins. Exploration covers the Minch and Outer Hebrides, including St Kilda. A wide variety of marine and bird life is to be viewed with the asistance of experienced guides.

SEAL ISLAND CRUISES
Fort William 01397 703919

Seal Island Cruises operates daily trips in season (Easter to October) to view seals on the local foreshores and islands. The experience provided is enjoyable, scenic and educational.

DUNVEGAN CASTLE SEAL TRIPS
Dunvegan, Isle of Skye 01470 521206

Daily trips to view wildlife off Skye's rugged coastline.

DONALD MACKINNON BOAT TRIPS (MV BELLA JANE)
Elgol, Broadford, Isle of Skye. 01471 86644

Sea watching is a speciality on these boat trips into Loch Scavaig under the towering Skye Cuillins.

SAIL GAIRLOCH
Tigh Na Bruaica, Gairloch, Ross-shire IV21 2BT

Operates both marine wildlife cruises and a centre displaying photographic and video coverage of local surveys of cetaceans, seal and seabird activity.

SCOURIE BOATS
Scourie IV27 4TE 01971 502011

Provides convenient 'pocket' tours, enabling all species of seabirds and other marine life to be seen. Visitors stay in the boat throughout the tour.

LAXFORD CRUISES
Seafood Restaurant, Tigh na Mara, Tarbet, By Scourie 01971 502251

This informative trip will include visits to a heronry and the nesting sites of cormorants and other sea birds. The boat is usually able to be within metres of grey and common seal colonies.

CAPE SEA TOURS
38 Sangomore, Durness, Sutherland 01971 511284

Cape Sea Tours offers a unique experience to enjoy the beautiful scenery and abundant wildlife of Durness.

STATESMAN CRUISES
66 Baddidaroch, Lochinver, Sutherland 01571 844446

A two-hour cruise up majestic Loch Glencoul to view Eas-Coul-Aulin Falls and an opportunity to view sea bird colonies and seals.

JOHN O' GROATS WILDLIFE CRUISES
Ferry Office, John O'Groats KW1 4YR 01955 611353

Boat trips into the Pentland Firth and round the great headlands of the Caithness coastline.

LAERLING Shetland
01595 693434

Described as "The ultimate Shetland wildlife cruise"

BRESSABOATS
Sundside, Bressay, Shetland ZE 2 9ER 01595 693434

Dunter II takes you closest to the stupendous sea cliffs of Noss. Skipper Dr. Jonathan Wills holds a boatmaster's license and has been boating and birdwatching around bressay and Noss for over 30 years.

FINDOCHTY WILDLIFE CRUISES
4 Craigview, Findochty, Moray AB56 2QF 01542 833867

Cruises to view dolphins and other marine life in the Moray Firth.

SEABOARD MARINE
The Boatyard, Nigg, by Tain, Rosshire 01862 871254

Wildlife watching, including dolphins, from the Cromarty Rose - the

small ferry plying the historic King's route at the entrance to the Moray Firth

MORAY FIRTH CRUISES
Shore Street Quay, Shore Street, Inverness 1V INF 01463 717900

Regular boat trips from Inverness Harbour into the Inner Moray Firth with a good chance of seeing dolphins and other wildlife.

MACAULEY CHARTERS
J. MacAulay, Harbour Offices, Longman Drive, Inverness 01463 717337

Wildlife cruises from Inverness harbour into the Inner Moray Firth looking for dolphins, sea birds and other marine life.

SCORPION
John Mackenzie, Carn Bhren, Portmahomack, Tain. 1V20 2YB 01862 871257

Boat trips into the unspoiled Dornoch Firth and the outer Moray Firth.

BENBOLA
K.Neilsen, 21 Great Eastern Road, Portessie, Buckie. AB56 1SR 01542 832289 or 0421 386803 (mobile)

This pleasure boat runs regular trips around the Moray Firth for the purpose of viewing the varied local wildlife and the beautiful coastline.

ANSTRUTHER PLEASURE TRIPS
7 Pittenweem Road, Anstruther Fife 01333 310103
Daily sailings from Anstruther to the bird and seal colonies of the Isle of May.

SEA WATCH FOUNDATION
7 Andrews Lane, South Water, West Sussex RH13 7DY 01403 733900

In conjunction with Western Isles Sailing, the Foundation conducts cetacean surveys every summer.

WESTERN ISLES SAILING & EXPLORATION
Coombe Down Cottage, Underhill Lane, Hassocks, West Sussex BN6 9PL 01273 846542

Specialists in wildlife cruises aboard an 80ft traditional wooden gaff ketch. Sailing out of Oban.

BASS ROCK
24 Victoria Road, North Berwick 01620 893863

Fred Marr takes parties of visitors to, and round, the Bass Rock, with its sea bird colonies, particularly thousands of gannets. Landing is at the discretion of the boatman and only after the breeding season has finished.

WILDLIFE AND VISITOR CENTRES
(where not referred to elsewhere)

THE SCOTTISH DEER CENTRE
The Bow-of-Fife, by Cupar KY15 4NQ 01337 810391

Scottish red deer in a parkland setting.

AN TAIRBEART
Tarbert, Loch Fyne, Argyll 01880 820190

The Centre's ranger, John Aiken, conducts a variety of activities, many of them with children in which the aim is to increase knowledge and understanding of the natural heritage and to help safeguard it for the future.

SKYE ENVIRONMENTAL CENTRE
Broadford, Isle of Skye IV49 9AQ 01471 822487

The Centre provides information on how to watch sensitively as well as where. It also manages a wild animal hospital.

ISLAY WILDLIFE INFORMATION CENTRE
Islay Natural History Trust, Port Charlotte, Isle of Islay PA49 7UN 01496 850288

The Information Centre provides information and helping experience about local wildlife and mounts exhibitions and displays.

ARNAMURCHAN NATURAL HISTORY CENTRE
Glenmore, Acharacle, Argyll PH36 4JG 01972 500254

The Natural History Centre at Glenmore provides the ideal introduction to the Peninsula's wildlife. The Living Building, built to attract as much wildlife as possible, offers you the chance to come face to face with birds and possibly even a pine marten.

Useful Contacts

There are many different organisations in Scotland concerned with wildlife and the countryside. Those listed below are confined to agencies, both government and voluntary, who hold and manage land for conservation purposes. Some of these organisations eg The Forestry Commission, have other duties, but the information given here relates only to their public access functions. A number of these bodies also have area offices in other parts of Scotland which will be able to provide more detailed information on particular localities.

Forestry Commission
231 Corstorphine Road,
Edinburgh EH12 7AT
Tel: 0131 334 0303

Provides public access to all state forests with special recreational facilities in forest parks.

John Muir Trust
12 Wellington Place, Leith,
Edinburgh EH6 7EQ
Tel: 0131 554 0114

Acquires property in the wilder areas of Scotland which are accessible to the public.

National Trust for Scotland
5 Charlotte Square, Edinburgh
EH2 4DU
Tel: 0131 226 5922

Apart from historic buildings and gardens, the trust also owns properties of wildlife and scenic interest where visitors are welcomed.

Royal Society for the Protection of Birds
12 Regent Terrace, Edinburgh EH7 5BN
Tel: 0131 557 6275

All of the Society's reserves are open to the public and most have special viewing facilities.

Scottish Natural Heritage
12 Hope Terrace, Edinburgh EH9 2AS
Tel: 0131 447 4784

The official government agency responsible for the establishment of National Nature Reserves, many of which are open to the public.

Scottish Wildlife Trust
Cramond House, Cramond Glebe Road, Edinburgh EH4 6NS
Tel: 0131 312 7765

A number of the reserves managed by this voluntary organisation have public access and information services.

The Woodland Trust
Glenruthven Mill, Abbey Road, Auchterarder Perthshire PH3 1DP
Tel: 01764 662554

Purchases and manages native woodlands and potential forest land for conservation and public enjoyment.

Further Reading

The Scenery of Scotland: The Structure Beneath W.J. Baird
National Museums of Scotland 1988

A short but excellent well-illustrated introduction to the geology and landforms of
Scotland for the non-specialist

A Guide to the Nature Reserves of Scotland ed. Linda Bennet
Macmillan 1989

A comprehensive guide with detailed descriptions of the more important protected
areas, although now slightly out of date.

Scottish Birds Valerie Thom
Collins Guide Harper Collins 1994

Good guide for those with little knowledge, clearly identifying the commoner species,
avoiding rarities or those very difficult to identify and with clear illustrations

Where to Watch Birds in Scotland (3rd edition) M. Madders and J.
Welstead
Christopher Helm, London 1997

Comprehensive guide to sites, with particularly useful notes on access, times to visit,
and local contacts.

Collins Field Guide to the Birds of Britain and Europe R. Petersen, G.
Mountfort & P.A.D. Hollom
Harper Collins 1993

The standard authoritative field guide with full details of all species and their distribution.

Scottish Wild Flowers Michael Scott
Collins Guide Harper Collins 1995

Concentrates on the commoner flowering plants well illustrated and with notes on
selected botanical sites

Scottish Wild Plants Philip Lusby and Jenny Wright
Royal Botanic Garden, Edinburgh The Stationery Office 1997

A beautifully illustrated publication on the history, ecology and conservation of 40
special native species, including many rare plants

The Natural Heritage of Scotland: An Overview
Scottish Natural Heritage 1995

A detailed factual summary of the present state of the natural resource of Scotland in
easy-to-digest form.

Scottish Wildlife - Animals   Ray Collier
Colin Baxter Photography Ltd. 1992

Excellent photography for identifying the animals of Scotland, accompanied by succinct text on their biology and habits.

The Nature of Scotland (2nd edition) Magnus Magnusson and Graham White
Canongate, Edinburgh 1997

The coffee-table photographic format belies the comprehensive information compiled by many experts on the landscape, wildlife and people related to natural heritage.

An Inhabited Solitude: Scotland - Land and People   James McCarthy
Luath Press, Edinburgh 1998

Focuses on the land uses which have influenced the natural history of Scotland with a view of the future for conservation.

Acknowledgements

I WOULD LIKE to acknowledge with sincere thanks a number of agencies and individuals who have helped in several ways. I am grateful to Magnus Magnusson for agreeing to write the foreword, more especially from his vantage point as the chairman of the Scotland's official conservation agency. Staff of that agency, Scottish Natural Heritage, have willingly provided up-to-date information, especially on species, and I have particularly appreciated the information given by Vincent Fleming. My thanks go also to staff of the Scottish Tourist Board for information and base maps, as well as helpful advice. I am pleased to record my acknowledgement here to the editors of SCENES (Scottish Environment News) for recent information on a wide range of conservation issues. Several of the non-government and voluntary conservation organisations helpfully provided site lists and details of particular locations, including RSPB, Scottish Wildlife Trust, John Muir Trust, and the Woodland Trust, National Trust for Scotland. I am indebted to A & M Consultants of Dunblane for assistance with access to the data base on wildlife tourism facilities prepared for the 1997 Report on Tourism and the Environment to the Tourism and Environment Task Force, to Duncan Bryden for his contribution of Sustainable Tourism, to Laurie Campbell for agreeing to contribute his incomparable photographic illustrations and advice on wildlife photography, and to Iain Sergeant for the atttractive line drawings. Rawden Goodier made helpful comments on the draft text. Lastly, but no means least, my appreciation must go to my wife not only for improving my aberrant syntax and punctuation, but also for excercising considerable forbearance throughout.

James McCarthy

Some other books published by **LUATH** PRESS

NATURAL SCOTLAND

An Inhabited Solitude: Scotland, Land and People

James McCarthy
ISBN 0 946487 30 8 PBK £6.99

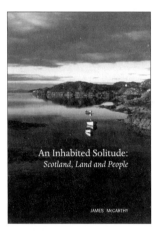

An Inhabited Solitude:
Scotland, Land and People

JAMES McCARTHY

'Scotland is the country above all others that I have seen, in which a man of imagination may carve out his own pleasures; there are so many inhabited solitudes.'

DOROTHY WORDSWORTH, in her journal of August 1803

In this informed and thought-provoking profile of Scotland's unique landscapes and the impact of humans on what we see now and in the future, James McCarthy leads us through the many aspects of the land and the people who inhabit it: natural Scotland; the rocks beneath; land ownership; the use of resources; people and place; conserving Scotland's heritage and much more.

Written in a highly readable style, this concise volume offers an understanding of the land as a whole. Emphasising the uniqueness of the Scottish environment, the author explores the links between this and other aspects of our culture as a key element in rediscovering a modern

sense of the Scottish identity and perception of nationhood.

'This book provides an engaging introduction to the mysteries of Scotland's people and landscapes. Difficult concepts are described in simple terms, providing the interested Scot or tourist with an invaluable overview of the country. The interdependent links between land and people are convincingly pointed up. Scotland's diverse culture, wildlife and landscapes are properly celebrated throughout. But it is in its final chapter that the book comes into its own. Here we are given a personal, thought provoking look at the nation today and its prospects in the next millennium. This is a book which will appeal to a wide audience. It fills an important niche which, to my knowledge, is filled by no other publications.'

BETSY KING, Chief Executive, Scottish Environmental Education Council.

Rum: Nature's Island

Magnus Magnusson KBE
ISBN 0 946487 32 4 £7.95 PBK

Rum: Nature's Island is the fascinating story of a Hebridean island from the earliest times through to the Clearances and its period as the sporting playground of a Lancashire industrial magnate, and on to its rebirth as a National Nature Reserve, a model for the active ecological management of Scotland's wild places.

Thoroughly researched and written in a lively accessible style, the book includes comprehensive coverage of the island's geology, animals and plants, and people, with a special chapter on the Edwardian extravaganza of Kinloch Castle. There is practical information for visitors to what was once known as 'the Forbidden Isle'; the book provides details of bothy and other accommodation, walks and nature trails. It closes with a positive vision for the island's future: biologically diverse, economically dynamic and ecologically sustainable.

Rum: Nature's Island is published in co-operation with Scottish Natural Heritage (of which Magnus Magnusson is Chairman) to mark the 40th anniversary of the acquisition of Rum by its predecessor, The Nature Conservancy.

LUATH GUIDES TO SCOTLAND

'Gentlemen, We have just returned from a six week stay in Scotland. I am convinced that Tom Atkinson is the best guidebook author I have ever read, about any place, any time.'
Edward Taylor, LOS ANGELES

These guides are not your traditional where-to-stay and what-to-eat books. They are companions in the rucksack or car seat, providing the discerning traveller with a blend of fiery opinion and moving description. Here you will find *'that curious pastiche of myths and legend and history that the Scots use to describe their heritage... what battle happened in which glen between which clans; where the Picts sacrificed bulls as recently as the 17th century... A lively counterpoint to the more standard, detached guidebook... Intriguing.'*
THE WASHINGTON POST

These are perfect guides for the discerning visitor or resident to keep close by for reading again and again, written by authors who invite you to share their intimate knowledge and love of the areas covered.

South West Scotland

Tom Atkinson

ISBN 0 946487 04 9 PBK £4.95

This descriptive guide to the magical country of Robert Burns covers Kyle, Carrick, Galloway, Dumfries-shire, Kirkcudbrightshire and Wigtownshire. Hills, unknown moors and unspoiled beaches grace a land steeped in history and legend and portrayed with affection and deep delight.

An essential book for the visitor who yearns to feel at home in this land of peace and grandeur.

The Lonely Lands

Tom Atkinson

ISBN 0 946487 10 3 PBK £4.95

A guide to Inveraray, Glencoe, Loch Awe, Loch Lomond, Cowal, the Kyles of Bute and all of central Argyll written with insight, sympathy and loving detail. Once Atkinson has taken you there, these lands can never feel lonely. 'I have sought to make the complex simple, the beautiful accessible and the strange familiar,' he writes, and indeed he brings to the land a knowledge and affection only accessible to someone with intimate knowledge of the area.

A must for travellers and natives who want to delve beneath the surface.

'Highly personal and somewhat quirky... steeped in the lore of Scotland.'
THE WASHINGTON POST

The Empty Lands

Tom Atkinson

ISBN 0 946487 13 8 PBK £4.95

The Highlands of Scotland from Ullapool to Bettyhill and Bonar Bridge to John O'Groats are landscapes of myth and legend, 'empty of people, but of nothing else that brings delight to any tired soul,' writes Atkinson. This highly personal guide describes Highland history and landscape with love, compassion and above all sheer magic.

Essential reading for anyone who has dreamed of the Highlands.

Roads to the Isles

Tom Atkinson

ISBN 0 946487 01 4 PBK £4.95

Ardnamurchan, Morvern, Morar, Moidart and the west coast to Ullapool are included in this guide to the Far West and Far North of Scotland. An unspoiled land of mountains, lochs and silver sands is brought to the walker's toe-tips (and to the reader's fingertips) in this stark, serene and evocative account of town, country and legend.

For any visitor to this Highland wonderland, Queen Victoria's favourite place on earth.

Highways and Byways in Mull and Iona

Peter Macnab
ISBN 0 946487 16 2 PBK £4.25

'The Isle of Mull is of Isles the fairest,
Of ocean's gems 'tis the first and rarest.'

So a local poet described it a hundred years ago, and this recently revised guide to Mull and sacred Iona, the most accessible islands of the Inner Hebrides, takes the reader on a delightful tour of these rare ocean gems, travelling with a native whose unparalleled knowledge and deep feeling for the area unlock the byways of the islands in all their natural beauty.

WALK WITH LUATH

Mountain Days & Bothy Nights

Dave Brown and Ian Mitchell
ISBN 0 946487 15 4 PBK £7.50

Acknowledged as a classic of mountain writing still in demand ten years after its first publication, this book takes you into the bothies, howffs and dosses on the Scottish hills. Fishgut Mac, Desperate Dan and Stumpy the Big Yin stalk hill and public house, evading gamekeepers and Royalty with a camaraderie which was the trademark of Scots hillwalking in the early days.

'The fun element comes through... how innocent the social polemic seems in our nastier world of today... the book for the rucksack this year.'
Hamish Brown, SCOTTISH MOUNTAINEERING CLUB JOURNAL

'The doings, sayings, incongruities and idiosyncrasies of the denizens of the bothy underworld... described in an easy philosophical style... an authentic word picture of this part of the climbing scene in latter-day Scotland, which, like any good picture, will increase in charm over the years.'
Iain Smart, SCOTTISH MOUNTAINEERING CLUB JOURNAL

The Joy of Hillwalking

Ralph Storer
ISBN 0 946487 28 6 PBK £7.50

Apart, perhaps, from the joy of sex, the joy of hillwalking brings more pleasure to more people than any other form of human activity.

'Alps, America, Scandinavia, you name it – Storer's been there, so why the hell shouldn't he bring all these various and varied places into his observations... [He] even admits to losing his virginity after a day on the Aggy Ridge... Well worth its place alongside Storer's earlier works.'
TAC

MUSIC AND DANCE

Highland Balls and Village Halls

GW Lockhart
ISBN 0 946487 12 X PBK £6.95

Acknowledged as a classic in Scottish dancing circles throughout the world. Anecdotes, Scottish history, dress and dance steps are all included in this

'delightful little book, full of interest... both a personal account and an understanding look at the making of traditions.'
NEW ZEALAND SCOTTISH COUNTRY DANCES MAGAZINE

'A delightful survey of Scottish dancing and custom. Informative, concise and opinionated, it guides the reader across the history and geography of country dance and ends by detailing the 12 dances every Scot should know – the most famous being the Eightsome Reel, "the greatest longest, rowdiest, most diabolically executed of all the Scottish country dances".'
THE HERALD

Fiddles & Folk

GW Lockhart
ISBN 0 946487 38 3 PBK £7.95

From Dougie MacLean, Hamish Henderson, the Battlefield Band, the Whistlebinkies, the Scottish Fiddle Orchestra, the McCalmans and many more come the stories that break down the musical barriers between Scotland's past and present, and between the diverse musical forms which have woven together to create the dynamism of the music today.

I have tried to avoid a formal approach to Scottish music as it affects those of us with our musical heritage coursing through our veins. The picture I have sought is one of many brush strokes, looking at how some individuals have come to the fore, examining their music, lives, thoughts, even philosophies.
WALLACE LOCKHART

For anyone whose heart lifts at the sound of fiddle or pipes, this book takes you on a delightful journey, full of humour and respect, in the company of some of the performers who have taken Scotland's music around the world and come back enriched.

...'I never had a narrow, woolly-jumper, fingers stuck in the ear approach to music. We have a musical heritage here that is the envy of the rest of the world. Most countries just can't compete,' Ian Green, of Greentrax Recordings says ...as young Scots tire of Oasis and Blur, they will realise that there is a wealth of young Scottish music on their doorstep just waiting to be discovered.

The Scotsman, March 1998

BIOGRAPHY
On the Trail of Robert Service
GW Lockhart
ISBN 0 946487 24 3 PBK £7.95

Known worldwide for his verses 'The Shooting of Dan McGrew' and 'The Cremation of Sam McGee', Service has woven his spell for Boy Scouts and learned professors alike. He chronicled the story of the Klondike Gold Rush, wandered the United States and Canada, Tahiti and Russia to become the bigger-than-life Bard of the Yukon. Whether you love or hate him, you can't ignore this cult figure. The book is a must for those who haven't yet met Robert Service.

'The story of a man who claimed that he wrote verse for those who wouldn't be seen dead reading poetry... this enthralling biography will delight Service lovers in both the Old World and the New.'
SCOTS INDEPENDENT

Bare Feet and Tackety Boots
Archie Cameron
ISBN 0 946487 17 0 PBK £7.95

The island of Rum before the First World War was the playground of its rich absentee landowner. A survivor of life a century gone tells his story. Factors and schoolmasters, midges and poaching, deer, ducks and MacBrayne's steamers: here social history and personal anecdote create a record of a way of life gone not long ago but already almost forgotten. This is the story the gentry couldn't tell.

'This book is an important piece of social history, for it gives an insight into how the other half lived in an era the likes of which will never be seen again'
FORTHRIGHT MAGAZINE

'The authentic breath of the pawky, country-wise estate employee.'
THE OBSERVER

'Well observed and detailed account of island life in the early years of this century.'
THE SCOTS MAGAZINE

'A very good read with the capacity to make the reader chuckle. A very talented writer.'
STORNOWAY GAZETTE

Come Dungeons Dark
John Taylor Caldwell
ISBN 0 946487 19 7 PBK £6.95

Glasgow anarchist Guy Aldred died with 10p in his pocket in 1963 claiming there was better company in Barlinnie Prison than in the Corridors of Power. 'The Red Scourge' is remembered here by one who worked with him and spent 27 years as part of his turbulent household, sparring with Lenin, Sylvia Pankhurst and others as he struggled for freedom for his beloved fellow-man.

'The welcome and long-awaited biography of... one of this country's most prolific radical propagandists... Crank or visionary?... whatever the verdict, the

Glasgow anarchist has finally been given a fitting memorial.'
THE SCOTSMAN

POETRY

Blind Harry's Wallace

William Hamilton of Gilbertfield
ISBN 0 946487 43 X HBK £15.00
ISBN 0 946487 33 2 PBK £7.50
The original story of the real braveheart, Sir William Wallace. Racy, blood on every page, violently anglophobic, grossly embellished, vulgar and disgusting, clumsy and stilted, a literary failure, a great epic.

Whatever the verdict on BLIND HARRY, this is the book which has done more than any other to frame the notion of Scotland's national identity. Despite its numerous 'historical inaccuracies', it remains the principal source for what we now know about the life of Wallace.

The novel and film *Braveheart* were based on the 1722 Hamilton edition of this epic poem. Burns, Wordsworth, Byron and others were greatly influenced by this version 'wherein the old obsolete words are rendered more intelligible', which is said to be the book, next to the Bible, most commonly found in Scottish households in the eighteenth century. Burns even admits to having 'borrowed... a couplet worthy of Homer' directly from Hamilton's version of BLIND HARRY to include in 'Scots wha hae'.

Elspeth King, in her introduction to this, the first accessible edition of BLIND HARRY in verse form since 1859, draws parallels between the situation in Scotland at the time of Wallace and that in Bosnia and Chechnya in the 1990s. Seven hundred years to the day after the Battle of Stirling Bridge, the 'Settled Will of the Scottish People' was expressed in the devolution referendum of 11 September 1997. She describes this as a landmark opportunity for mature reflection on how the nation has been shaped, and sees BLIND HARRY'S WALLACE as an essential and compelling text for this purpose.

'Builder of the literary foundations of a national hero-cult in a free and powerful country.'
ALEXANDER STODDART, sculptor

'A true bard of the people.'
TOM SCOTT, THE PENGUIN BOOK OF SCOTTISH VERSE, on Blind Harry.

'A more inventive writer than Shakespeare.'
RANDALL WALLACE

'The story of Wallace poured a Scottish prejudice in my veins which will boil along until the floodgates of life shut in eternal rest.'
ROBERT BURNS

'Hamilton's couplets are not the best poetry you will ever read, but they rattle along at a fair pace. In re-issuing this work, the publishers have re-opened the spring from which most of our conceptions of the Wallace legend come.'
SCOTLAND ON SUNDAY

'The return of Blind Harry's Wallace, a man who makes Mel look like a wimp.'
THE SCOTSMAN

Poems to be Read Aloud

selected and introduced by Tom Atkinson
ISBN 0 946487 00 6 PBK £5.00
This personal collection of doggerel and verse ranging from the tear-jerking 'Green Eye of the Yellow God' to the rarely-printed bawdy 'Eskimo Nell' has a lively cult following. Much borrowed and rarely returned, this is a book for reading aloud in very good company, preferably after a dram or twa. You are guaranteed a warm welcome if you arrive at a gathering with this little volume in your pocket.

'The essence is the audience.'
Tom Atkinson

FOLKLORE

The Supernatural Highlands

Francis Thompson
ISBN 0 946487 31 6 PBK £8.99
An authoritative exploration of the otherworld of the Highlander, happenings and beings hitherto thought to be outwith the ordinary forces of nature. A simple intro-

duction to the way of life of rural Highland and Island communities, this new edition weaves a path through second sight, the evil eye, witchcraft, ghosts, fairies and other supernatural beings, offering new sight-lines on areas of belief once dismissed as folklore and superstition.

WALKS & SHORT WALKS

The highly respected and continually updated guides to the Cairngorms.

'Particularly good on local wildlife and how to see it.'
THE COUNTRYMAN

Walks in the Cairngorms

Ernest Cross

ISBN 0 946487 09 X PBK £3.95

This selection of walks celebrates the rare birds, animals, plants and geological wonders of a region often believed difficult to penetrate on foot. Nothing is difficult with this guide in your pocket, as Cross gives a choice for every walker, and includes valuable tips on mountain safety and weather advice.

Ideal for walkers of all ages and skiers waiting for snowier skies.

Short Walks in the Cairngorms

Ernest Cross

ISBN 0 946487 23 5 PBK £3.95

Cross wrote this volume after overhearing a walker remark that there were no short walks for lazy ramblers in the Cairngorm region. Here is the answer: rambles through scenic woods with a welcoming pub at the end, birdwatching hints, glacier holes, or for the fit and ambitious, scrambles up hills to admire vistas of glorious scenery. Wildlife in the Cairngorms is unequalled elsewhere in Britain, and here it is brought to the binoculars of any walker who treads quietly and with respect.

HISTORY

The Bannockburn Years

William Scott

ISBN 0 946487 34 0 PBK £7.95

This is a love story, at the time of Scotland's greatest success.

How did the Scots win the War of Independence, against a neighbour ten times as powerful?

Did the Scots have a secret weapon at their disposal?

Was the involvement of women a deciding factor?

Should Scotland now become independent?

These questions lie at the heart of the medieval manuscript by John Bannatyne of Bute, genius, commander of the Scottish archers at Bannockburn, and eye-witness of Robert Bruce's heroic leadership.

A present day solicitor is asked to stand for the independence party in an election. In a client's will, he stumbles across reference to a manuscript of value to the Nation State of Scotland which he tracks down to the Island of Bute.

Is the document authentic? In the course of his investigation, involving a WWII fighter-pilot, the solicitor also discovers his answer to the question: should Scotland be independent now?

Written with pace and passion, William Scott has devised an original vehicle for looking at the future of Scotland. He presents, for the first time, a convincing explanation of how the victory at Bannockburn was achieved, with a rigorous examination of the history as part of the story.

Winner of the 1997 Constable Trophy, the premier award in Scotland for an unpublished novel, this book offers new insigts to both the general and academic reader, sure to provoke further discussion and debate.

'... a brilliant storyteller ... I shall expect to see your name writ large hereafter.'
NIGEL TRANTER, October 1997

SOCIAL HISTORY

The Crofting Years

Francis Thompson

ISBN 0 946487 06 5 PBK £6.95

Crofting is much more than a way of life. It is a storehouse of cultural, linguistic and moral values which holds together a scattered and struggling rural population. This book fills a blank in the written history of crofting over the last two centuries. Bloody conflicts and gunboat diplomacy, treachery, compassion, music and story: all figure in this mine of information on crofting in the Highlands and Islands of Scotland.

'I would recommend this book to all who are interested in the past, but even more so to those who are interested in the future survival of our way of life and culture.'
STORNOWAY GAZETTE

Luath Press Limited

committed to publishing well written books worth reading

LUATH PRESS takes its name from Robert Burns, whose little collie Luath (*Gael.*, swift or nimble) tripped up Jean Armour at a wedding and gave him the chance to speak to the woman who was to be his wife and the abiding love of his life. Burns called one of *The Twa Dogs* Luath after Cuchullin's hunting dog in Ossian's Fingal. Luath Press grew up in the heart of Burns country, and now resides a few steps up the road from Burns' first lodgings in Edinburgh's Royal Mile.

Luath offers you distinctive writing with a hint of unexpected pleasures.

Most UK bookshops either carry our books in stock or can order them for you. To order direct from us, please send a £sterling cheque, postal order, international money order or your credit card details (number, address of cardholder and expiry date) to us at the address below. Please add post and packing as follows: UK – £1.00 per delivery address; overseas surface mail – £2.50 per delivery address; overseas airmail – £3.50 for the first book to each delivery address, plus £1.00 for each additional book by airmail to the same address. If your order is a gift, we will happily enclose your card or message at no extra charge.

Luath Press Limited

543/2 Castlehill
The Royal Mile
Edinburgh EH1 2ND

Telephone: 0131 225 4326
Fax: 0131 225 4324
email: gavin.macdougall@luath.co.uk
Website: www.luath.co.uk